SATPREM was born in 1923 in Paris. After a seafaring childhood in Brittany, he was arrested by the Gestapo at the age of twenty and was imprisoned for a year and a half in German concentration camps. Deeply broken in his heart and soul and body, he traveled to Egypt, then to India, where he held his first and last position as a civil servant in the government of Pondicherry. There, he discovered the "new evolution" envisioned by Sri Aurobindo: "Man is a transitional being," handed in his resignation and left for South America, spending a year in the heart of the Amazon jungle (the basis for his first novel, *The Gold Washer*), then moving on to Brazil and Africa, ever in search of the "true adventure."

He returned to India in 1953, at the age of thirty, became a mendicant Sannyasi, practiced Tantrism (the basis for his second novel, *By the Body of the Earth*), and finally abandoned all these paths to put himself at the service of Mother and Sri Aurobindo, to whom he dedicated his first nonfiction work, *Sri Aurobindo or The Adventure of Consciousness,* then a second nonfiction work, *On the Way to Supermanhood*. He stayed with Mother for 19 years, becoming her confidant and witness and collecting numerous personal conversations that form *Mother's Agenda* (13 volumes of which 6 are published in English). This adventure with her who was seeking the secret of the transition to the next species gave rise to a trilogy on Mother (*The Divine Materialism, The New Species, The Mutation of Death*), then to a fable, *Gringo*, set in the jungle, like his first novel 20 years before, and finally to *The Mind of the Cells,* his latest nonfiction work, which distills the essence of Mother's discovery: a change in the genetic program and a different view of death. Satprem now lives withdrawn from all public and literary life to devote himself entirely to trying to implement in his own body and cells the "new way of being on earth."

Mother
or
The Mutation of Death

## ALSO BY SATPREM

*Sri Aurobindo*
*or The Adventure of Consciousness* (1974)
*By the Body of the Earth* (1978)
*Mother:*
1. *The Divine Materialism* (1980)
2. *The New Species* (1982)
3. *The Mutation of Death* (1987)
*The Mind of the Cells* (1982)
*On the Way to Supermanhood* (1986)

---

### *Mother's Agenda*
1951-1973
13 volumes

Recorded by Satprem in the course of numerous personal conversations with Mother, the complete logbook of her fabulous exploration in the cellular consciousness of the human body. Twenty-three years of experiences which parallel some of the most recent theories of modern physics. Perhaps the key to man's passage to the next species. (Vols. 1, 2, 3, 4, 12 & 13 published in English)

# Mother
## or
# The Mutation of Death

*by* Satprem

Translated from the French

Institute for Evolutionary Research
200 Park Avenue, New York, N.Y. 10166

This book was originally published in France under the title *Mère ou la mutation de la mort* by Éditions Robert Laffont, S.A., Paris. © Satprem 1976.

*Mother or The Mutation of Death.* English translation copyright © by Institute for Evolutionary Research, Ltd. All rights reserved. For information address:
    *U.S.A. and Europe:* Institute for Evolutionary Research
    200 Park Avenue, Suite 303 East, New York, N.Y. 10166 U.S.A.

    *India and Asia*: Mira Aditi Center
    Aspiration, Auroville, Kottakuppam 605104 (T.N.) INDIA

Library of Congress Cataloging-in-Publication Data

---

Satprem, 1923-
    Mother, or, The mutation of death
    Translation of: Mère ou la mutation de la mort
    Bibliography: p.
    Contents: v. 1. The divine materialism -- v. 2. The new species -- v. 3. The mutation of death.
    1. Mother, 1878-1973. 2. Spiritual life (Hinduism)
I. Title.
BL1175.M43S2713 1987    294.5'5 [B]    86-27467
ISBN 0-938710-17-6

---

Manufactured in the United States of America
*First Edition*

## Contents

*Part One* — THE CELL WITHOUT CODE

1. Transforming Matter or Negating Matter? 1
2. A Body Without Memory .......... 7
   *The Horrible Thing* 8
   *A New Mind* 12
3. The Mind of the Cells ........... 19
   *The Fundamental Cage* 20
   *The Agent of Mutation* 22
4. The Mutation of Death .......... 29
   *Stealth* 31
5. The Traveler ................. 36
6. The Cellular Spinning .......... 40
   *Cellular Plasticity* 41
7. A Willing Automatism .......... 46
   *Direct Functioning* 47
   *A Smile That Knows Everything* 51
8. The Exit from the Second Web ...... 55
   *A Way of Being* 57
   *The Exit from the Second Web* 59
   *The Key to the New Body* 65
   *Primary Matter* 69
   *The Mystery of the Unknown* 75

| 9 | Innumerable Life . . . . . . . . . . . . . . . . 79 |
|---|---|
| | *The Son of the Cells  81* |
| | *Cellular Ubiquity  83* |
| | *The Fact of the Corpse  89* |

| 10 | Victory over Death or in Death? . . . . . . 92 |
|---|---|

| 11 | Transformation . . . . . . . . . . . . . . . . 98 |
|---|---|
| | *Transformation or Change of Perspective?  99* |
| | *The Problem of Transformation  105* |
| | *Cellular Stabilization  109* |
| | *Cellular Survival  111* |

| 12 | The Permeation . . . . . . . . . . . . . . 116 |
|---|---|
| | *Iridescent Materiality  118* |
| | *The Supramental Invasion  121* |

| 13 | Cellular Time . . . . . . . . . . . . . . . . 128 |
|---|---|
| | *The Central Experience  131* |
| | *The Change of Position or Massive Time  137* |
| | *The Contradiction  143* |
| | *A Cataleptic Trance?  146* |
| | *The Innumerable Present  150* |

*Part Two* — A DANGEROUS . . . UNKNOWN

| 14 | The Residue . . . . . . . . . . . . . . . . 159 |
|---|---|
| | *A Question of Patience  167* |

| 15 | The Supreme Door . . . . . . . . . . . . 172 |
|---|---|
| | *The Old Way Is Dying  173* |
| | *The Mystery of the Contradiction  178* |
| | *The Layer of Carbon  185* |

| 16 | Overlife . . . . . . . . . . . . . . . . . . 192 |
|---|---|
| | *Two States of Matter* 193 |
| | *Tomorrow's Unknown* 199 |
| 17 | Uninterrupted Physical Life . . . . . . . 208 |
| | *Another Physical Air* 210 |
| | *Another Physical Rhythm* 213 |
| | *At Any Second* 218 |
| 18 | The Problem of the World . . . . . . . . 221 |
| | *The Triple Acceleration* 222 |
| | *More and More, More and More . . .* 228 |
| | *Ineluctable Victory* 230 |
| | *The Negation* 236 |
| 19 | The Impossible Solution . . . . . . . . . 244 |
| | *On the Threshold of a Great Secret* 244 |
| | *The Great Immobility* 249 |
| | *As Inside an Egg* 254 |
| | *Sleeping Beauty* 259 |
| 20 | The Last Path . . . . . . . . . . . . . . . 267 |
| | *The Last Meeting* 268 |
| | *Hold On* 273 |
| | *The Most Beautiful Fairy Tale* 278 |
| | Works of the Mother . . . . . . . . . . . 288 |
| | Works of Sri Aurobindo . . . . . . . . . 289 |
| | References . . . . . . . . . . . . . . . . . 292 |

To
> Her

> May our aspiration have
> the power to reveal
> what is hidden
> and to manifest the unexpected.
>                               *S.*

*Part One*

THE CELL WITHOUT CODE

> *Her single will*
> *opposed the cosmic rule.*[1]
> —Sri Aurobindo

ONE

# Transforming Matter or Negating Matter?

If that true Vibration, that little lining of Truth, were to replace the vibration of falsehood, we would find ourselves in an enormously changed world—inconceivably changed, because we cannot conceive of what is simple. We can conceive of fairies, of gods and all sorts of grandiose complications or supercontrivances, and in fact we spend our time inventing complications to simplify our complications, but that which does not need any contrivance, that which flows as naturally as spring water, well ... A spring is fresh at each instant. It is something that finds its way according to its own inclination of truth, and that opens its way by the simple power of being what it is. To be is *to be capable* of being what one is at every second: an apple tree, a doe, one kind of song or another—and it just sings, that's all. To be a man, this human transition, was exactly to be capable of being what one is not, and since that is not really possible, it meant an unreal power within a fortress of unreality—the only difference is that now the unreality is flying in our face. But what about that real world, suddenly real, stripped of the illusion—clear? What about those not yet suffocated beings who suddenly open transparent eyes—by the millions? It is quite dizzying. It is dreadfully miraculous! It may also be very funny, but what is it? A world where, suddenly, everything communicates—for that was what the fortress was all about: nothing communicated. A world where we know all there is to know at every second, and exactly in the right amount, like the bird, simply. We know all that we do not know, because the fortress is what made the wall of unknowing. To start with, all schools crumble. Only the immense School of the Game of Life is left, and maybe schools of physical education for the body, or, rather, education of the *consciousness* of the body. There is no more cramming of heads, because there is no more fortress to cram: the whole world is an open book. And then each one is what he is; that makes for many different kinds of music. And since it is no longer necessary to rob one's neighbor in order to fill one's own cellar, no longer necessary to earn a false living in order to try to make a real life for oneself on the side, no longer necessary to be other than what one is, all sense of competition crumbles: What is the sense in "succeeding" to be like one's neighbor? Each person and each nation, and each group, if one still needs to be in a group, needs nothing other than to be what he or it is quite joyfully, because to be is to feel the joy of what one is, purely, without addition or subtraction. And without frontiers. There's nothing to earn! —except ourselves, more and more beautifully, more and more limpidly, and more and more powerfully. For we can also do all that we cannot do—the fortress is what made a wall of powerlessness (very wisely so, in fact, or automatically, we could say, because we would have seized hold of that Power to break our neighbor's neck and falsify everything, as usual). But there, in that world stripped of illusion, there is no need of morality or police or courts of law anymore; there is just the automatic power of what one *is*, and of naturally accomplishing what one is. How could there be fakers in that clear world? If they could exist at all, they would be extraordinarily visible, all twisted like their thoughts, wrapped in grayness and blackness like rats. There is no sense pretending; everything is obvious and clear, and out you go! But what is truly marvelous is that, *by himself,* the faker will no longer be possible—those poor fakers who so deceive themselves, who are so full of illusions, who hunger after everything, who suffer and struggle to get what they are not ... but they can *no longer* deceive themselves! That's it, a world where one can no longer deceive oneself. What is the point of painting one's face like a clown, then? The only means of power left is *to be,* more and more. Who would choose to have cancer, then—the real cancer, the one of Falsehood, the one that causes all the little cancers in the flesh?

Of course, a world without doctors, without lawyers, without—we have a long list of complications. No more telephones, because we communicate with everything—it was a walled fortress. No more distance, no more separation, consciousness is everywhere—it was a fortress of unconsciousness. The web has torn open, and we run—*that* runs everywhere. No more trepidation or hurry—hurry for what? Tomorrow is perfectly today and each second is perfectly what it must be, fresh, like the little spring. An amazingly simple world ... like the truth. Just a veil tearing. A few million clear eyes suddenly staring and pressing together against the veil.

The ghosts will say that this is not possible because, for them, everything is "not-possible," except the cage that creates all their powers—evangelical, governmental, scientific, constitutional and eternal. All in the the same bag. They are the evangelists of death, so naturally they want to keep it! But there is still one point. A mortal point—this body. And this point of Death is like the very key to Falsehood, or to the Truth concealed by that Falsehood—for in fact only Truth exists in this world, and even Falsehood would not exist if there were not some Truth behind it. This is the great simulacrum of Falsehood stuck on an imperturbable Truth. Unstick the Falsehood, and the real earth emerges. It can be tomorrow. It is not very far; it does not take "time" to happen. All it takes is for all those little consciousnesses to reach the "point of homogeneity," as almost happened in 1968, but without the knowledge of the process, or of the reason or Power behind it. The Amazon is right here, radiant, funny, unveiled. A formidable, worldwide breath of air. But this old body is here—this old body which has undergone the transition up to this point, this old animal residue. What will happen to it? Mother had come exactly to that point. For her, everything was unveiled, without limits, without possibility of illness, for illness is only the material, bodily expression of Falsehood. She was in the real earth in advance. She was preparing the real earth for everyone by wearing down the web in her body. Illnesses are eliminated, wear and tear itself can be eliminated. There is no more "friction" or trepidation in those cells of the clear world. But they are nevertheless animal cells. And then again, what is the sense of staying in a ninety-year-old body? And even if we conceive—as is very conceivable—that younger bodies, even very young, might effect the transition and break through the web at eighteen, what is the meaning of a body that has to eat, digest, carry weight? It seems that its very system of functioning, even pure, even freed from diseases, contains its own seed of death and decay: as you eat so must you be eaten. The body seems to be the very symbol of Death.

And what does "death" mean when there is no more "other side"? Indeed, the other side side is right here, once we have gone through the web. What happens then? We enter our "body of consciousness," as it were, the very one that is our lining, our reality, our body of reality, the one that Mother saw more distinctly than the bones and flesh of the fakers, which shone or faded away depending on the quality of consciousness. It is an old story. We have always had a body of consciousness—we even come back life after life to make that body grow, to develop it, universalize it, beautify it . . . teach it to love. For the cage is the site of Love—the site of suffering. This is perhaps the great secret of the cage. We come back here again and again until we have learned to love all and be all—in short, to be divine. There are not many men who do their true work in the cage, but there are some. The fakers vanish; they do not have a body on the "other side"; they are just an agglomeration of matter, which simply dissolves. But what happens to the others at the end of the cycle of growth, when their true body, the body of consciousness, is completely formed, developed, conscious, loving? Do they leave the animal rag and disappear, disencumbered, into the real earth—"ghosts" in reverse? One can well imagine a world where all the conscious bodies gambol happily *on earth,* the real earth, while the ghosts of the "obverse side" occupy the front stage, the false earth—and the two worlds remain as if superimposed, without connection. It is *already* happening. It is where Sri Aurobindo and many conscious beings live. But that does not seem to be the evolutionary solution. If we were formed in matter, it means that matter itself holds its own plenitude and ultimate fulfillment—where is the seed that yields a non-tree? Therefore, this seed of matter, symbolized by the body, must have its own meaning and its own key.

The death of the body must hold the key to its transformation.

A butterfly coming out of a caterpillar, yes, but in a *material* body, on the "obverse side" of the world.

Or is this "obverse side" definitely false and this matter definitely false? Then one simply leaves it—it is the cage's matter, and one simply goes off and flutters elsewhere-here.

There does remain a corpse, however, a symbol of death. And can a fully conscious and fully truthful being end up as a corpse, even if that corpse is false and he continues capering about in his other body? Truth cannot end up as Falsehood.

Death, the corpse, must be the ultimate key.

Something has to happen *there*.

The key to death.

The obstacle, the Negation *must* be the means of the passage to another state *in matter*.

Or else matter is meaningless and let's all go and be charming little ghosts . . . if we can.

What is the secret of death?

What is the ultimate secret of matter?

This is the mystery of Mother's last five years.

*A dangerous . . . unknown,* She said.

We will perhaps transform that last cage that the body is only when we discover the absolute Love concealed behind that absolute pain. Then we will discover that matter is the site of absolute Love. Death can only be transformed into its absolute opposite of Love. That is the reason this cage was invented. It has been matter's very quest since the first atomic fire.

And it is the ultimate transfiguring Fire.

TWO

# A Body Without Memory

On August 22, 1968, I received a short note from Mother. I had not seen her since August 10. The heart is giving out, the pulse is "more than erratic." Just the same, She had appeared on her balcony, upstairs, like the poop deck of a great ship, on the 15th of August, for Sri Aurobindo's birthday. She was all wrapped in her silver cape, so pale. She remained on her feet for five minutes. There were two assistants behind her to keep her from falling. And the crowd below. I was then reminded of that little story about Queen Elizabeth the First, who, despite the doctors' protests, tore herself from her deathbed to receive a delegation of merchants: "We shall die afterwards." It is Mother who told me that story, and it is Mother to a T. That little note of August 22 is very typical: *Here are some soups. You must be hungry* [there were several packets of powdered soup]. *This time, it is truly interesting—but rather total and radical. But how far, far away we are from the goal. ... I will try to remember.* One may be dying, but it is very interesting just the same—a subject for study. Mother would have made a perfect physicist—though this was actually the new physics. And then, don't forget to eat—when *She* could no longer eat anything!

## The Horrible Thing

Indeed, it was "rather total and radical." She was sitting in her very low rosewood armchair, which would be her seat right up to the end, always facing west, toward Sri Aurobindo under the great copper-pod tree. There was a small cushion under her sandaled feet. The armchair was covered with pale yellow Bangalore silk. There was a fragrance of "Muguet de Mai" in the air, her favorite perfume, shipped straight from Provence (the "Power of Purity," as She called it). She looked strangely diaphanous, and her voice especially had changed a lot. It sounded more and more like that of a child. I had never thought that She might die, and in fact I never thought She could die. The intensity of the shock was due to the sheer swiftness of the process: ten years compressed into a week. *I must work fast, you see.... They all think it's the end.* "No, no! Absolutely not!" I protested. "No, we all have faith that this is truly the ultimate possibility, and that it *cannot* not work." *Do they understand?* "They know that work is being accomplished." *Yes? All right, then.* And She laughed without believing a word of it.

A radical operation, the exact repetition of the 1962 turning point, but to a more total and definitive degree:

> *The mind and vital sent packing*
> *so the physical is left alone*
> *to fend for itself.*

And She handed me an illegible note scribbled in pencil. In other words, the experience of the body, pure, completely on its own. *If you prefer, apparently, I had become an idiot. I didn't know anything anymore.* She no longer saw, no longer heard, no longer knew how to do anything, even walk—the forgetfulness of everything. And yet, "something" was there to make her body still operate, act, coordinate everything—still speak with crystalline (but very particular) intelligence. She spoke to me till the end, and her mumblings were like drops of pure light, sometimes overwhelming with power. That "something" was what was under study. What is left when everything has been removed. The pure body. And not a single ounce of vital energy left—an almost impotent body—while there was that cataract of staggering power all around her. . . . A surprising contradiction. But these last five years are full of jolting contradictions—jolting, because in that impossible living paradox that She was incarnating more and more, one seemed, sometimes, to grasp something so new . . . that it was almost unbelievable. As if the earth gave way under one's feet, but not into an empty hole: into an unbelievable—there is no other word for it—Marvel. No more mind, no more vital strength—She would never recover them. They were gone forever. But She would eventually move quite well, resume walking, writing, receiving one to two hundred people a day. She would even exercise every day with an oculist's chart to learn to see in our way again—an indomitable will. But there was another law. There was a "something else." Another possibility secretly, invisibly, but irresistibly growing in that totally annulled body—which could grow only because it was annulled.

Hell. Five years of hell.

*It is truly hell. That Possibility is the only thing that keeps it from being hell. It is because, behind this hell, there is that Possibility—which is living, real, existing, which one can touch, in which one can live—otherwise it's . . . hellish. You see, in ordinary men, it seems as if all the states of being* [reflexes, feelings, instincts, thoughts, ideals, etc.] *have been beaten together, you know, like mayonnaise! All the states are all whipped together in the greatest confusion, and hence "the horrible thing" can be borne because of everything else that's mixed with it! But if you separate . . . oh!* Remove feelings, thoughts, automatic reactions, memories and, of course, all possible ideals, all tastes, all mental constructions from A to Z, from top to bottom ... and what is left is "the horrible thing." Mother would remain in the horrible thing, pure, till the end.

A new being is made of cells free of all programming. Free of mental programming, vital programming, material programming. And what can possibly come out of such a state, of that horrible nothing?

An impossible state.

A completely unlivable state.

But it could be only *because* it was unlivable.

Pure matter, we could say.

And universal matter into the bargain. Not an iota of protection against the world's tide, the world's thoughts, the world's reactions, the world's diseases.... Sometimes her little childlike moans could be heard right down in the Ashram courtyard. It was heartrending. And She apologized. *One sees and hears that CLAMOR of protest, of grief, of suffering—that clamor is all over the earth—and these cells are a little ashamed.... I spend almost entire days and nights in silence, but seeing, seeing....* There were no more thoughts, nothing, only images, an immense and constant living film—one screen, another screen—which She entered alive, here, there and everywhere, to experience this call for help, that illness, this murder, that mean or petty act.... Everything was living and lived. A flood of pain: the world's pain. *I don't have the feeling or perception of being a separate individual. I just have countless experiences, by the dozen, demonstrating that the identification or unity with other bodies is what makes me feel the misery of this person, the misery of that person, the misery—why, EVERYTHING is misery! I mean, it isn't a selfish complaint.* She seemed to apologize. Then She looked at all this earth before her, or within her: *There is a very clear and spontaneous sense that it's impossible to put aside a small portion of the whole and make it into something harmonious so long as everything isn't. But why,*

*why, why? ... This physical is really a mystery. I understand the people who said: It has to be abolished; it's falsehood—but it is NOT true. It is NOT a falsehood. It's—it's what? To call it a "distortion" doesn't mean anything. For instance, when I am told that somebody is ill, more than 99 times out of a hundred I already have experienced the same thing. I have experienced it as if it were part of my own physical being—an immense physical being, you know, immense and without any precise form. So ...* And I replied to Mother, "Well, that means that the consciousness of THE WHOLE must change. It's always the same problem: once the whole has progressed, changed its consciousness, then the 'material fact' should become different." *It would seem to be so. There is no escaping, no dividing that. The individuality is just a means of action for transforming the whole—how I understand those who said that one had to escape! It demands such a change ... it requires almost an eternity of time.* "One cannot be transformed without everything being transformed," I continued. "That is, 'one' speeds up the transformation of the whole." *Yes, that's it.* And, as if on reflection, She added, *It is quite obvious that if it were not unbearable, it would never change. And since it is unbearable, well, you really feel like running away—but it's impossible, you know! It's just their childishness to think that one can get away from it: it's NOT POSSIBLE. That delays the outcome.*

She lived in that growing "not possible" as if She were in the very heart of the pounding of the world. One *cannot* get away; there is *no* "other side"—so everything must always be begun again, painfully, one Christ after another. But something else *can* come out of it, by virtue of that impossibility. One does not get out of evolution, ever, whatever side one may be on, living or supposedly dead, but a new being can come out of the rubble of the old matter. That's all. That's the only possibility. For who would say that our celebrated chromosomes can invent a new being? *Whose* chromosomes? Those of our atavistic clutter? Of a mad lottery?

Maybe for the first time on earth, at least on the earth of men, there was an aggregate of human matter free from all its genetic memories—except for the great living-lived memory of human Pain: the "horrible thing," pure. And somewhere in there, under the Pressure of that pain, something ...

*Truly interesting,* She concluded. And She laughed every time She could, because laughter was the best way to rout Death.

## A New Mind

There were still other little notes, scribbled in pencil during those nights of August 1968, "to try to remember." During those great turning points or difficulties, even when She could not see me, Mother always turned her beam toward me, remembered me, as if She knew in advance the hour of separation when She would have to build a fragile bridge over the unknown and attempt to link yesterday's earth to tomorrow's. Here is what those little broken, overlapping lines were saying:

> *For several hours, the landscapes*
> *were marvelous,*
> *of a perfect harmony.*
> *Constant visions.*
> *Each thing has a reason, a precise goal*
> *to express nonmental states of consciousness.*
> *Landscapes.*
> *Structures.*
> *Cities.*
> *All immense and very diverse,*

> *encompassing the whole visual field and*
> *translating states of consciousness*
> *of the body.*
> *Many, many structures,*
> *immense cities under construction. . . .*

Yes, the world under construction, the future world under construction. *I didn't hear, I didn't see, I didn't speak anymore; I lived in this all the time, all the time, all the time, night and day. A body without mind and vital.* There were only those perceptions: it lived in the soul's perceptions. There were the soul's perceptions of others, the soul's perceptions of the earth, the soul's perceptions—the soul's perceptions translated into images. In that pure, annulled matter—inconceivable for us, for it would not even be like a baby's matter—there only remained the pain of the world, and images. They were the only means of perception left. But not just images that are seen—images that are lived. *Those scenes of the soul's perceptions—there were so many things . . . so many marvels! Absolutely no mental notion can be as marvelous as that—none. I spent moments . . . All that we can feel or see humanly is nothing compared to that. I spent the most . . . marvelous hours—I think the most marvelous hours one can have on earth. But without any thought, without any thought.* And they were *physical,* material "images" of the earth. Not "visions"—the real earth. We perhaps do not know how beautiful the earth is. Mother did talk of a "more effective," "fuller" vision. *But it was not "seen" as one sees a picture; it was being IN IT, being in a certain place. I have never seen or felt anything as beautiful as that, and it was not "felt"; it was—I don't know how to explain it. There were absolutely marvelous and unique moments. And it was not thought. I couldn't even start to describe it—describe how? You can only describe when you start to think.* This is perhaps how the next instrument of the new being will work: things are no longer thought; they are lived. Each thing unfolds its own complete landscape which explains everything. Mother was trying to explain how the new being works. *The mind and vital were instruments for churning matter—churning, churning in every possible way* [to awaken matter to the existence of its own buried consciousness]: *the vital through sensations, the mind through thoughts. But to me they look like transitory instruments which will be replaced by other states of consciousness. They are a stage of universal growth, and they will fall away like instruments that are no longer needed.* The evolutionary atrophy of old limbs grown useless.

And Mother pointed to another note:

> *The state of consciousness of the body*
> *and the quality of its activity depend*
> *on the individual, or individuals, it is with....*

*Ah, that was very interesting! It was very interesting because I could see it all change. Someone would approach me, and it would all change. Something would happen to somebody, and it would all change....* Each minute, the "landscape" of each individual—the attendants, those who served Mother—changed, said certain things, automatically expressed what was happening, with all the depth, the colors, the "decors," the close or distant ramifications, in the most minute detail—all our thoughts about the world and all our "real" eyes are like one-dimensional photographs of an unsuspected *earthly* reality. *And then, when the slightest thing changed in their consciousness, it would all start changing! It was a sort of perpetual kaleidoscope, day and night. If only there had been a way of recording it, it would have been marvelous...* [matter's eyes]. *And the body was in the midst of it all, you know, almost porous—porous, without resistance—as if everything went right through it.* No more barriers. A state of instantaneous, exact, ubiquitous, marvelous—and painful—consciousness. A marvel and a hell simultaneously. Like the obverse and reverse of the earth experienced together. One and the *same* earth.

And a "porous" body.

What actually takes place in such a body? How can it live, function, organize itself? Yes, this was perhaps a new kind of baby upon earth. But one still must eat in the old way, sustain oneself in the old way, listen to and experience stupidities in the old way, and pain in the old way—keep a coherent link with the old world, live in an old body in the depths of which something . . . was beginning to stir. Something that looked like a new body in formation. And this is where we begin to touch a very concrete, practical phenomenon, which, despite its almost insignificant simplicity, is the greatest earthly turning point since the appearance of the first protozoan. A new life form. Such is the experience that is gradually going to unveil itself, without our realizing at first exactly what it is, so insignificant is it. Was the little protozoan so significant in its pond?

> *The body, the cells of the body*
> *would like to be in contact with the true being*
> *without going through the vital*
> *or even the mind.*
> *That is what is happening.*

And all of a sudden, Mother remembered: *Oh, I realized that the cells, everywhere, constantly, kept repeating the mantra: OM Namo Bhagavate, OM Namo Bhagavate. . . . Constantly, constantly.*

> *OM Namo Bhagavate is being repeated*
> *spontaneously and automatically,*
> *in a sort of hazy peace.*

It was the last little scribbled note.

A body, a new body. An awakened and conscious cellular matter, completely fresh to the world, free of laws and programming, groping within a great Consciousness, repeating and repeating the mantra, refashioning a new way of being for itself—a "way," that is, a form. Can one imagine a baby growing up in an old body, at the very cellular root? Something that no longer produces enzymes and catastrophes, something utterly free of programs—unprogrammed—something filled with the constructive energy of a new shoot—can one conceive of a young shoot coming out of the frozen earth? Crack, and the ground breaks open! Something unthinkable. But it is a fact. That fact would unfold for five years before our eyes—painfully, paradoxically. Indeed, the old earth must be broken. The old strength is there no longer; the old, tyrannical, but convenient mental coordinator of all that bodily organization is there no longer—so what remains? What does a body do when the laws for standing up have been removed? A porous body. What are those pure, virgin cells going to produce?

And this is where something very simple but of enormous —perhaps devastating—consequence began to develop in matter: a new mind. A mind of the body, of the cells. Something that began to organize itself and furrow matter while repeating the mantra—but furrows that no longer formed a cage, no longer went round in a circle, no longer became encrusted in a molecule of amino acid. Something congruent with that "vertical time" in which each thing that takes place is new at each second, without trace, without "imprint," without "memory," and yet having a sort of memory of the future, as we could say with Sri Aurobindo, a memory working ahead instead of behind—no longer a little granary collecting one grain after another, growing larger and subjecting you more and more to its accumulated, stereotyped push: a pull, not a push. The mighty pull toward the Future at each second. A mind that would be exactly the opposite of the physical mind and would take its place—which was beginning to take its place in Mother's body. Another Vibration replaces the mortal little vibration that is panic-stricken by life's onrush and tries to shut itself up, to barricade itself, and at the same time to store up, to hoard the food it can no longer pluck from life's great flow, to freeze all movements in an immovable and invulnerable pattern, to sleep and sleep in order to duplicate the peace of the stone, to die in order to be done once and for all with that toil— a trepidant mind which is only a tissue of repetitions and crystallized habits, which even succeeded in crystallizing matter into a certain form of being as if *it* were the vehicle and the driver, the true agent of crystallization of matter into a form, the very base or deep vibration of each molecule of DNA and amino acid. A new mantra of matter replaces the old, maleficent mantra of the physical mind which repeats and repeats its mortal refrain. A new crystallizing agent of matter. We say that the molecules of protein are what imparts form to bodies— a giraffe, a mouse, a man—but this is the superficial, external translation, the material *garb* of a particular quality of vibration. If the vibration changes, the type of crystallization or materialization changes. If the physical mind, which gathers and perpetuates this mode of vibration, changes, matter's entire organization must change.

It is the phenomenon that took place in Mother's body.

Cells that spontaneously repeat the mantra.

A new mind in matter, or of matter, like a new driving force and the agent of a new type of body on earth, which will not necessarily be crystallized in the same way.

It is not a new way of thinking that is developed in matter; it is a new way of being of matter.

It is the embryo of the new species.

The key to transforming the body.

For ten years, from 1958 to 1968, She had worked to get through the web, to purify the cells of their old programming, the spell of the physical mind, and now the Work was to build a new type of body in matter.

THREE

# The Mind of the Cells

Actually, this flowering of the mind of the cells, this dramatic development in Mother's yoga (and Sri Aurobindo's)—so dramatic that in 1971, three years later, Mother would say, *But it's radical, my child! You can't imagine. It's like—I could really say that I've become a new person* (a new person at ninety-three, after undergoing countless experiences that others would regard as summits)—this simple, so simple and so fragile a thing astir in her body, that mantra repeating itself on its own, is the real knot of the battle Mother had fought since 1958—ten years. Not that that mind of the cells had not tried, often, to supplant the physical mind. But each time it was as though resubmerged, or, at best, it passively submitted to the higher mind, to the spiritual Vision, like a baby rather overwhelmed by its parents of Light. Even when unrestrained by the sordid physical mind, it turned to the "higher" Light, in awe, to "do its best"—but that best was worth nothing, or at least could continue forever to grind out a higher, sanctified and purified humanity, all the old improved evolutionary jumble. It took the radical cleanup of 1968 for those cells to be freed from the "best" of the higher mind as well as the "worst" of the physical mind—for them to be purely themselves. And this is where the miracle truly began. This is the secret that each "man" of the next species will have to discover—which will probably be easier to discover now that one body has uncovered it and understood the incredible power and incredible freedom that lay there.

## The Fundamental Cage

We talk of the "mind of the cells" and we talk of the "intellectual mind," the "intuitive mind," the "liberated mind" in its heaven high above, but there is only one Mind, and the word may not be appropriate. These are one and the same consciousness, one and the same power with different vibratory modes or in different "mediums." Sri Aurobindo even said that that "Mind," or Consciousness-Force, can be found in each atom and particle. That Consciousness, or Power, has always circulated freely through matter—in fact, it is the very constituent of matter, the Energy put into equation by Einstein (what has not been put into equation is the consciousness of that Energy). With each species developing in evolution, from the small gelatinous blob to the hominid, that Power concretized itself, formed into one way of being or another, one vibratory habit or another, a certain molecule that fixed its circuit and seemed to want to repeat forever the acquired habit, except when, by "miracle" or "mutation" (both words are equally empty), the habit was broken at one point or another and a new species, or new type, began spinning and repeating a new habit or a new vibratory mode of the same eternal Power. Our "molecules" and "mutations" explain nothing. They are the outer shell of the phenomenon. We can go on forever peeling molecules and particles, and we will always find something else, which conceals something else, which conceals something else. ... What we will never grasp is the internal tension of the Consciousness that produces that break and not any other, at that precise moment and not any other. We can batter all the molecules like the laboratory demiurges that we are, but we will never produce what follows man. It is as simple as that. We will not gain man's secret. This is the true limitation of our science. Likewise, we can batter and bombard all the particles we like, but we will not master the great Energy, except to produce deadly devices. We will not gain matter's secret. But at the human level of the evolution of that great Power, a sort of black veil fell, a second cage over the first fundamental, physiological cage that is proper to each species. And what if our "accident" is also the key to breaking the fundamental cage. Instead of letting the Power spin as usual as in other mammals and follow the little course of the simple cellular mind that keeps repeating its seasons, its marriages, its urges and impulses amid an open milieu where everything communicates, "feels," "responds" and lives in a harmony that may very well seem divine to us compared to our grating life, we have built the cage of the ego—the I—and have taken on individual twists in a shell of noncommunication with the all rest. Hence all the complications, distortions, suffocations, fears, etc., which have slowly formed the leaden cage of the physical mind since the Pleistocene, to the point that the cellular mind could not even spin its simple, healthy habits without the other one interfering and terrorizing, medicalizing, hypnotizing and traumatizing everything with all the contrivances made necessary by its confinement and all the false habits of its individuality in constant competition with everything else. A mongoose does not have a physical mind; man has one, infernally. It is our great misery. It is the enormous web that sinks deep into human matter, so much so that we feel we cannot get rid of it without getting rid of the very life of our body.

It is this cage of the "I" that all philosophies, all religions, all sociologies come up against; it is to remedy it or break it that we produce Marxism, heavens and hells, democracies and telephones, but the real exit—the evolutionary door—is down below, as we have seen, behind or beneath the web of the physical mind, in the mind of the cells. And if we broke that cage, we would probably rediscover the freedom and happy harmony of the animal—and many other things that Mother discovered while going through the web. There is found the site of the integral unity that is sought in vain by Marxism, religions and our equations. This is Einstein's "unified field." But first, evolution never goes backward: we will not return to the mongoose—and even then, the fundamental cage would still remain. That digesting and aging structure. And this is where one perfectly grasps the evolutionary stratagem that has caused us to undergo this painful transition through the "I" of the supercage. One day or another, compelled by our very suffocation to get out of the web, we find ourselves—and by now with an individual view, an individual understanding of the evolutionary program—before that primary, cellular substance free of its old ghosts, and we discover things that no animal could have discovered because it is perfectly comfortable in its cage. Our worst was our supreme best in hiding.

## The Agent of Mutation

Mother's discoveries spanned many years. They started well before 1968. But it is only when the last higher remnants of the mind were swept away that those cells were suddenly left on their own, without control, except for whatever could arise from their own depths. From then on arose a formidable Power—the very one that drove the atoms and little men, or great men, saints and sages, and all the evolutionary circuits, but through one filter, two filters, layer upon layer of filters. And it was right there, pure, whole. All-powerful. *A power that can crush everything to bits and rebuild everything.* It was in 1964 that Mother for the first time touched that cellular foundation, and what is most remarkable is that there is no more "personal" experience at that level—of course, since there is no more person, just the entire world!—and when one has an experience, it is as if (not as if: it is) the *whole* terrestrial field had the experience, were put into contact with it. This is what Mother said in 1964: *Something has begun to descend—not "descend," but manifest, penetrate—penetrate and fill this terrestrial consciousness. It had such a force, such a power!—an intensity I had never felt before in matter—such a* STABILITY, *such a power! It was all power and pressure to move ahead—toward progress, evolution, transformation. It was the joy of progress. . . . Amidst all that, in that mass of experience, as if standing out from the rest, was a perception of the gorilla, of the enormous thrust for progress that will make a man out of him. It was very strange. There was a fantastic physical power, coupled with an intense joy of progress* [that joy of the cells, always; it is the main attribute of the cellular consciousness], *of forging ahead, and there was a sort of simian form advancing toward manhood. Then, it's as if the same thing repeated itself in the evolutionary spiral: the same brute Power, the same vital energy (without any comparison, for man has completely lost that), the same formidable energy of life in the animal was coming back into the human consciousness, but along with everything that has been brought by the evolution of the Mind (which accounted for a rather painful detour) and was* CHANGED *into the light of a higher certitude and peace. And this was not something that came and faded and came again. No, it was . . . an immensity, a solid, full,* ESTABLISHED *immensity. Not something that comes, presents itself and says: this is how it will be—it was* THERE. *And it was not something being infused into the Mind; it was an infusion into Life, into the terrestrial, material substance, which had become alive. Even plants partook of the experience. It was not something restricted to the mental being; the whole vital, material substance of the earth received that joy of the power of progress—it was triumphant, absolutely triumphant.... A diamond-like sparkle. When I got up this morning, I had the feeling that a corner was turned FOR THE EARTH. That people do not realize it does not matter at all.*

Like Sri Aurobindo: it explains itself.

And Mother added this: *In any case, the experience has been decisive in the sense that it has coordinated all those scattered little promises, those scattered little progresses* [the hundreds of little experiences that sprouted up on every side while She proceeded blindly in the Forest]. *And there was a rather clear perception that, soon, the state of being or way of being (I think it is called "modus vivendi") of the body, of this piece of terrestrial matter, could be altered, directed, entirely controlled by the direct Will* [the great total Consciousness]. *Because it was as if all the illusions had been destroyed one after another* [the illusion of illness, of death, of decay, of the instinct for survival—all the indisputable physiological illusions that stick to us in the web of the physical mind], *and each time an illusion was dispelled, it produced one of those little promises, which came one after another, as forerunners of something that was still to come.* It is only four years later, in 1968, that this great "direct Will" would take charge of the cells without going through the filters of the mind, the vital or even the most "spiritual" filters in the world.

But not only does one have to be able to withstand the Power—this is the whole preparation, widening, universalization, impersonalization that Mother's body underwent—something has to hold it there, as it were, some kind of turbine or condenser, a fixative element that keeps it from escaping like air. What can possibly "condense," fix the Power in those cells? It was the very first question Mother asked herself in 1958. But in 1965 came a double discovery, negative and positive, very profound and yet very simple (there is nothing spectacular at the cellular level; it is the mind that loves and thrives on complications), which is truly the prelude, the key to the new species or, rather, to the change of species. The agent of mutation. For a mutation to take place in that cellular substance, something must mutate, something must stop absorbing the current as usual and eternally churning out the same vibrations. Until now, the physical mind had functioned as the condenser, and it had a peculiar habit of "condensing" catastrophes and steering the current according to the old medical and "reasonable" atavistic patterns (it is remarkable how "faithfully" it reproduced *Merck's Manual,* and it makes you wonder whether the disease followed Merck or Merck followed the disease), and Mother had tried and tried to silence it, to annul it. But She realized this, which is her first, negative discovery: *It was very difficult to get rid of, because it was so intimately tied to the physical substance of the body and its present form.... It was difficult, and whenever I tried and a more profound consciousness tried to manifest, it caused fainting. I mean that uniting, merging, identifying with the supreme Presence* [or the Power, the "other thing," the great Current] *without that, WITHOUT THAT PHYSICAL MIND, while it was annulled, caused fainting.* In other words, the physical mind was a sort of link, the nexus between matter (or the body) and the Power that drives and animates matter. Remove it, and nothing retains the Current anymore; it just goes right through—one faints. Hence, it was like a death penalty for the human system: one cannot get rid of that old fabricator of catastrophes, for without it everything falls to pieces. No mutation is possible—it will go on churning out catastrophes to the death. There is no other nexus.

But in 1965, on the 21st of July, to be exact, there took place a microscopic little phenomenon, like a whisper, which changes everything.

From beneath that sort of fossilized crust of the physical mind something suddenly emerged, a crack, a breach in the carapace, and another voice, a new whisper in the body: *There's a slight hope that that material mind, that mind of the cells, could change—all of a sudden, here it is offering up a prayer. A prayer ... you know, the way I used to in "Prayers and Meditations"* [Mother's former journal at the beginning of the century]. *That was the mind offering up prayers (it had experiences and it offered up prayers); well, here, now the cells are having the experience: a very intense aspiration, and all of a sudden it all begins to express itself in words. It's as if that mind of the body uttered a prayer, in the name of the body (as if the body was beginning to become "mentalized"). And it has a deep sense of the oneness of matter. So there was the sense of all of matter—all terrestrial human matter—and it said:*

> *The other states of being, the mind, the vital,*
> *may be satisfied with intermediary contacts* ....

That is, with having a relationship with all the intermediary states of being: the gods, etc., the heavens, illuminations, revelations and music of all kinds.

> *Only the supreme Lord*
> *can satisfy me.*

*And, all at once, there was such a clear vision that only what is supremely perfect can bring plenitude to this body. I found that interesting. It's the beginning of something. It began with a sense of disgust—a sickening disgust—for all these miseries, these weaknesses, these strains, this malaise, all this conflict and friction, oh! ... And it was very interesting, because there was that disgust and, at the same time, a sort of suggestion of Annihilation, of Nothingness—of eternal Peace* [the physical mind's great aspiration: reverting to the stillness of the stone]. *And it swept all that away* [it is the cellular mind itself that does the sweeping], *as if the body drew itself up: "Eh, but wait* a minute! This is not what I want. I want . . . (and there was a dazzling light, a dazzling golden light) . . . I want the splendor of your Consciousness."* The first pure reaction of the cellular consciousness. *I have the feeling that it's like grasping the tail of the solution. A whole world is beginning to open up. We'll see.*

The mind of the cells itself reacting against the catastrophic hypnotism of the physical mind and whispering a "prayer"—a prayer, that is, a verbalization, a vibration. The first pure vibration of the cells.

It was the new nexus of Power—the mind of the cells.

The shattering of the old pattern—the agent of mutation.

Something that was going to absorb the pure Current.

Just a little whisper within the cells. A first mantra of matter.

As if matter were born anew.

And such an aspiration, such an intense joy of aspiration in those cells, like a golden breath in the depths of matter: *Everything mental seems cold and dry, yes, dry, lifeless—it's luminous, pretty, pleasant, but it's cold, lifeless. While this aspiration here, oh, it has a power—a power of realization— that is absolutely extraordinary! If it organizes itself, it will be possible to accomplish something. There is a lot of power there.*

It is the very power that drives the atom, the pure supramental power, the Energy of all mutations and all transformations of matter: "A power that can crush everything to bits and rebuild everything."

Now, the "new language" of the cells had to be learned, this new mind of matter had to be "organized," the golden vibration had to be repeated until it could transmute the old crust.

But the key had been found.

It is the key to getting through the second web, the one that binds us to the body of a mammal as it bound others to an aquatic or reptilian body. We wonder how we poor human beings can go through that Wall when Mother and Sri

Aurobindo had to display so much heroism and tenacity to effect the passage, but the way is open. We now know where the key is—the whole difficulty is not in the difficulty itself; it is in not knowing how to take the difficulty. This is the whole work of the evolutionary pioneers: to find the way. Now we know: the cells have to start repeating a mantra, and they will continue repeating it all by themselves until the old machine comes apart. This is how the web will be cleared away and how the cells will be purified, automatically. We must learn to drive a mantra into the body. Then it will repeat it as conscientiously as "I forgot to turn off the electricity," or "I'm going to get cancer," or ... It will repeat it like a mule, twenty-four hours a day. It is as simple as that. As far-reaching a discovery as the operation of a little mental vibration in a few gray cells.

It may well be Sri Aurobindo's "mathematical formula."

FOUR

# The Mutation of Death

Great revolutions are always simple.
We obsessively believe that we have to devise great schemes and revolutionize the world and effect spectacular changes, but we revolutionize nothing; there is no mutation. We merely recombine the same elements in a different order, and since those elements are of no value in any order, we constantly end up with what could be called an improvement of the catastrophe. A true revolution, a true mutation means that a tiny new element has managed to sneak in and alters the value of each of the old elements. It is not a change of order; it is a change of value. And what was worthless or adverse in any order suddenly takes on a new sign, as if it simply had not found its own key and was adverse because it had not found its own key—and ultimately nothing was adverse, not a single atom, but everything was waiting for the little key. And this is why we will find nothing, change nothing, revolutionize nothing so long as we do not find the fundamental key—because it will change all the signs. And what, finally, must we find? What is there to uncover in this enormous universe, with us in it? Let's be simple. Joy, of course, which is love. And what is the opposite of that simple reason for being? Death, of course, which is non-joy and non-reason for being. And if it has no reason for being,

it follows that it should not be, that it is the unreality of the universe—yet it is the one thing that suffocates us most, that has a bearing on all other signs and almost nullifies them. There are not a hundred solutions; there is only one. Not a hundred revolutions, only one. A single mutation of death. Then everything will change, automatically. All the other signs will take on a different value.

And chances are, the little key will be found with that which we have to mutate, or in it—at the origin of death. And what is it that dies? Cells die, because they keep spinning out a song of death, an unreal song, a non-reason for being. But this is not true! Life *cannot* spin non-life; it can only spin joy and more joy, because it is joy itself. It only spins death because it has not found the little key, or in order to force us to find the little key to reality, real life. It is as simple as this: what causes death may cause life *as well*. It is simply a key turned in the wrong direction. It is not something to remove. There is nothing to remove from the universe—to put where, into what dumping ground outside of the universe? It is something to turn in a different direction, and everything else will turn in a different direction. *We are searching for the process so as to have the power of undoing what has been done,* She said. To undo death. *After all these years, there is something in me that would like to have the power or the key—how it works. And should one not feel or experience or see (but "see" in an ACTIVE way) how it all became twisted* [Mother turned her hand in one direction] *in order to be able to do this* [She turned her hand in the opposite direction]? To experience what causes death in order to turn it in the other direction. To experience death.

Maybe this is what She was going to do for five years . . . and after.

And She added this: *What's interesting, now that the mind of the cells has organized itself, is that, at lightning speed, it seems to go through the whole process of human mental develop-*

*ment again in order to reach . . . just the key.* The key to death. *There is, of course, the sense that the state we are in is a deceptive unreality, but there is a sort of need or aspiration to find, not the mental or moral "why" of it, nothing of the kind, but the HOW: how it became twisted like this* [Mother turned her hand again] *in order to put it straight like that.* The cells are the key to spinning in one direction or the other. But there is no death *in fact.* There is no *fact* of death; there is a wrong spinning. Death is not a cellular phenomenon; it is a cellular absurdity. It is unreality stuck onto a reality we have not yet found. Once that reality is touched—a reality of joy—death will become unreal all by itself. Death is not a reality of matter; it is a particular unreality of matter—it dies because it is not what it is. *Every time I ask my body what IT would like, all the cells say: No no! We are immortal. We want to be immortal. We are not tired; we are ready to struggle for centuries if necessary—we were created for immortality and we want immortality. And in fact, I realize (I don't think this is anything unique and exceptional) that the closer you get to the cell, the more the cell says: Why, I'm immortal!*

This is the cellular *fact.* The only real fact.

The reality of the cells is immortal life.

Perhaps even modern biology would not disagree.

There is nothing that is death in the substance of the cells.

The joy of a pure little cell holds the mutation of death.

## Stealth

Such is the great, simple revolution that took place from 1968 onward: all of a sudden, the cells, pure, left to their own working, began to spin in the other direction, as if it were the most natural thing in the world—it was indeed the only naturalness in the world. They began to unravel death, quite spontaneously and naturally. There are no words to describe that simple little wonder because it is so simple, so inconspicuous—and yet it is what will change the whole world, whether it wants it or not, realizes it or not now. It is what has begun on earth. The mutation of death *has* begun. For a cell is not a separate entity; it runs through everything. There is one single body. And I watched the tiny marvel develop, hesitating on Mother's lips—sometimes I even thought I had caught it in my own body: *People have experiences, and they don't know it!* She said. Because it is not mental, because we do not look in the right place. We only understand an "experience" when we have mentalized it, "understood," whereas this takes place without our understanding and without any need for understanding—it takes place constantly! The world is changing course without its knowledge. It does not notice the little golden breath that is simply unraveling death, as if nothing were happening. Death does not "take place," so no one notices anything! One has to die for real to notice the damned thing. For who can perceive those microscopic little deaths that are spun and spun and end up causing the real death? And that little breath that prevents it from taking place? For it is truly like an imperceptible little golden breath that circulates all by itself, without any interference—when we do not interfere, it happens *all by itself,* that's the beauty of it! Whenever suggestions of death or annihilation or eternal peace came into contact with the cells, they simply rejected them, like that: Away with you, I don't want you! It is that simple. Whenever suggestions of illness approached: Away with you, I don't want you! Whenever suggestions of old age came: That's a lie, I don't want you! And everything was like a world of constant evil suggestions, everywhere, truly as if we splashed around in a sea of filth without realizing it, soaking it up like pure air—away with you, I don't want you! It is the cells *themselves* that throw them out. But one has to be somewhat free from the mental racket to notice it, because it is so quiet, without violence, crystalline, weightless like the cells themselves, and as obviously powerful as the child for whom all this simply does not exist, is not—puff, I just blow on you, you don't smell very good! This is not me. *I had told you about that "morbid imagination" of the body—it's completely gone, finished, all cleared up! From the moment the body reacted by saying: No, this is disgusting; what on earth is all this!—it's gone. That is what is so remarkable with this body: vitally and mentally, things have to be repeated again and again to establish the experience; the body is less prompt in opening up, but once it has understood or has had the proper experience, it's done. It's ESTABLISHED. That's what is so remarkable. And it's very quiet. And whenever certain things try to come back (even when they are at some distance, only at the periphery), it says: Ah, no, no! I don't want this anymore; it belongs to the past. IT is the one that does the work, because of the change that took place in that mind.* It is matter itself making its own revolution.

This is a phenomenon we ourselves might have noticed when we were a little "clear": we sense a suggestion approaching at some distance, at the "periphery," as Mother says, maybe a few feet from the body, a grating suggestion, like a tiny wave with a particular odor, or a suggestion of accident, a sexual suggestion, a suggestion of headache—nothing but suggestions; we live in a world of suggestions!– and suddenly we feel a sort of expansion in the cells of the body, yes, like something gorging itself with sun or light and creating a warm intensity (it's strange, it has almost the quality of love), a very compact and at the same time clear and light vibration, which rises from within, all by itself, without any interference or call on our part, without fuss, as simple as can be—poof, the suggestion is *automatically* dispelled! It no longer exists. A golden expansion. With the mind, we can fight ten times, a hundred times, keep the suggestions at arm's length to prevent their entry, but the minute we relax the arm, it's over, they come right in, and then we really have to struggle to get rid of them, or else catch a real fever. But here, there is *nothing* to do! It is automatic and radical. Just a little golden breath. The cells do the work.

And if one has a mantra, it becomes formidably effective.

And they do this constantly, everywhere—wherever there is a particle of sincere good will. It is like a constant cleaning out of death: those thousands and billions of sly suggestions that make a corpse in the end, a cancer in the end, an unbelievable waste, when, in truth, in *fact,* there was just a single little golden song that yearned to spin out the joy and beauty of life. So we wonder what is going to happen. Because it is happening in the body of people, in the body of nations, in the body of the earth—a great cleaning out. There is, of course, the heaviness that is growing heavier and heavier, darker and darker, almost visible to the naked eye, as if it were swallowing a double or triple dose of death, frantically spinning out its little mortal trepidation—but it is gorging itself with its own death; the unreality is becoming fantastic, almost phantasmagoric. But it is a total unreality. There is not a whisper of life in it. It is an enormous bubble painted steel-gray, a bogeyman inflated to the size of the earth—an overblown air bag. And underneath is that quiet, imperceptible but *imperturbable* little golden vibration which cleans and cleans, until not a single root of death is left standing, just that bubble over our heads. Then the bodily substance will be clear, and the bubble will collapse at one stroke—poof! ... It doesn't exist. It did not exist. Maybe there will not even be any "good guys" and "bad guys" in all this (still our human childishness). The bodily substance is totally *good.* The whites as well as the yellows will find themselves cleaned in their bodies without realizing it, and when everything is all clean and spotless underneath, they will look at their air bag *uncomprehendingly.* They will not believe their eyes. Then it will really collapse out of bewilderment.

## THE MUTATION OF DEATH

And now we seem so clearly to understand this often-quoted passage by Sri Aurobindo, including a mysterious little line that nobody really understood then:

> *When darkness deepens strangling the earth's breast*
> *AND MAN'S CORPOREAL MIND IS THE ONLY LAMP,*
> *As a thief's in the night shall be the covert tread*
> *Of one who steps unseen into his house.*
> *A Voice ill-heard shall speak, the soul obey,*
> *A power into mind's inner chamber steal,*
> *A charm and sweetness open life's closed doors*
> *And beauty conquer the resisting world,*
> *The truth-light capture Nature by surprise,*
> *A stealth of God compel the heart to bliss*
> *And earth grow unexpectedly divine.*
> *In Matter shall be lit the spirit's glow,*
> *In body and body kindled the sacred birth;* . . .
> *A few shall see what none yet understands;*
> *God shall grow up while the wise men talk and sleep;*
> *For man shall not know the coming till its hour*
> *And belief shall be not till the work is done.*[1]

Shall we let the phenomenon take place without paying a little attention . . . and without vibrating a little in tune with that golden sweetness that runs, as if inadvertently, underneath our unreal monsters?

The mutation of death is today.

FIVE

# The Traveler

But the fundamental cage would still remain.
Even with that mortal spinning abolished, along with all its causes for illness, distortion and misery—all that human pain which is a kind of radical ill-health—even with life prolonged at will, as it were, by the elimination of all friction and the plasticity of the great universal Rhythm, there would still remain that which fundamentally causes death: the death of bird and beast and all species up until now. Indeed, death will remain so long as the evolutionary design is not fulfilled, because it is the evolutionary means of progress. If the evolutionary design were to evolve an enhanced, harmonious and clash-free humanity, Death would probably disappear once that goal is attained, since there would no longer be any reason to die. But evolution does make the little birds die. Hence, there is something more, or other, to discover. Death still remains the door to the secret, the unmistakable sign that the one and only evolutionary Secret has not been found. And from one stage to another we are led ever closer, ever more into the heart of that fundamental matter which dies because it cannot, at least not yet, do what it wants. Because it has not yet found itself. And ultimately, this body, which seems to us an instrument, a mere material prop to enable us to gambol intelligently on the good

earth's surface, might be in fact the very crux of the story. A living heart of matter and of each cell of matter and of each atom. A secret original principle that appears only at the end, when the whole Amazon and every portion of the Amazon is uncovered. We believed that the little mental man had to be the explorer and final discoverer, but it could be that the "something" that created the first matter so many thousands of millions of years ago also created ultimate matter and that the traveler of this long journey was there with the first atom and the first molecule, or rather was *inside* that first atom and first molecule. We are knocking at the door of the ancient traveler. Each death knocks at his door; each life knocks at his door. We will die so long as we do not find that traveler.

Matter remains therefore our enigma.

What is in it?

No philosophy, of course, no religion either, thank "God"! What, then?

The closer one draws to the cell, the more mysterious it becomes—or more marvelous, we could say, but it is a marvel so "other," so yawning, as it were, that it is a little frightening. It is really the great "gap," or great evolutionary rift through which one suddenly lands in another country—not even "country": another being within us. At any rate, a "something" that is bewildering for all our physiological structures, which unremittingly spin their little enzymes, little membranes and exquisite, ready-made molecules. All of a sudden, nothing is "ready-made" anymore. It no longer works, or it works entirely differently. We are touching the very root of the fundamental cage, the one that made delightful little birds and that could make delightful little men, if such were the evolutionary goal. But it appears that we got through a first web in order to get through a second one. It would seem that we were not created to be exquisite prisoners after all, even of a molecule. We go on knocking at the door of death.

Death is perhaps the traveler's last mask.

For, after all, let us be simple, starkly simple: What is the real point of this wretched, or not so wretched, evolutionary story in which we are the more or less consenting and painful pawns? What is the meaning of all this? To end up as superbirds, which will eventually grow tired of their small wings or big wings? To create marvelous and ubiquitous supermen, who will be glutted with ubiquity, sated with marvels? Enough! To create enchanting lands dotted with fairy-like, multicolored superforests and seas unlike any other? To create, always to create something else. But so long as it is something else, it will never be *that*.

That, full.

That, simple, right here, without plus or minus.

That, which fills one at each second without needing anything else.

That, like breathing, and that's all.

That, like loving, and forever.

That.

That, yes. For if that is missing, we have nothing, and thousands of earthly paradises or not piled on top of one another will not fill that want.

Perhaps this is what the traveler is.

Perhaps the traveler is love.

We are knocking at the door of love, and so long as we do not find it, we will find nothing. This is perhaps what matter's secret is. And when we have come in contact with it in each cell and each atom, the doors of death will open up—it was so simple.

It is Simplicity.

That is why a little atom caught fire one day.

It is to experience that at each second and in everything that we took on this shell, this mantle of feathers, this mantle of man, but *that* was there all the time, within—and now it wants to be what it is. What will be *its* mantle?... What is this mantle of matter? The rag one discards after use, or the very body of the traveler—an unknown body, but one that must have its means of fabrication in the cells; the traveler is not going to fall from the sky; he has always been on earth with us, crawling with us, walking with us, struggling with us. And now what?

Where are you, traveler? What are you going to do with your little cells, which have toiled so long? Food for death, or another, inconceivable life, another body to be made?

A body of your love, which will be our only love.

And everything will be full.

And everything will be simple.

It will be.

Deep in the great eyes of the future, we see a golden flame rising.

SIX

# The Cellular Spinning

A whole new world reveals itself at the level of the cells, a very surprising and unexpected world, which seems so removed from the knowledge of the microscope and the biologist's rigid laws! It looks almost like the difference between seeing Guiana's forests from the air and walking the same forest on foot along the smooth banks of the Oyapock river. For this is *experiencing* the cellular level. That's the whole difference. And those laws that appeared so inflexible, those chains and spirals of DNA that seemed to be the key, quite simply look like the usual coagulations floating and whirling on a great river and, well, following that current because it happens to run that way. We have mistaken the whirling of the little flotsam for the law of the great Current. It is true that the stars that float in the great sea above, crossing each other, meeting or moving apart, for a moment seem to determine the character of a man or a people, because everything responds and corresponds—the little flotsam as well as the faint star and the molecule—but there is a great Current carrying everything. Instead of entering a cellular prison, one discovers an extraordinarily supple and fluid world—open. We have imprisoned everything; we are the great prisoners of our heads. And quite surprisingly, the nice little molecule follows the law of the dictionary, and it can continue following it for thousands and millions of years because nothing is more obliging than a molecule, until some upheaval of Nature makes a rip in its routine or simply somebody a little less stupid says to it: But why don't you turn in another direction? And it quite simply starts turning in another direction—but there has to be a little "I want" in the heart of the molecule. It may be that we have traveled all that distance just to discover that "I want." Our journey is a journey of freedom, and love at the end, for pure love can only exist in absolute freedom. In fact, we are prisoners of our laws only insofar as we *need* to be prisoners—they are little evolutionary confines for the babies of the earth—and all evolution tends to create new needs that will naturally shatter the old law. So long as there is no need, it is hopeless; it just keeps on turning and turning. Perhaps this is the moment of the Great Need? But to take the laws as laws is the eternal human folly.

## Cellular Plasticity

I therefore expected to see Mother struggle with millennial cellular imperatives, a sort of physiological and ... genial impossibility, for after all even the cellular Don Quixotes have their worth in a world shriveled up from head to toe by legalities. But not at all! The problem or difficulty does not lie where we think it does. What appears implacable to us is child's play—it is what does not appear implacable that is terrible! *What takes the most time is to become conscious of what must be changed,* She said, *to have a conscious contact that enables it to change.* A contact. Step by step, Mother went from one discovery to another—one could almost say that there is nothing to "discover," just to remove the walls that prevent us from seeing what is there. Illusion upon illusion to be disillusioned before we can arrive at reality. Mother is the great destroyer of illusions. With her, one breathed in a world that was finally possible, while everything is usually not-possible-not-possible-not-possible. How we can live in that without suffocating is a perpetual mystery. But the final, joyous suffocation that will break all those little legal windmills is approaching. It is only a matter of reaching the right dose. The moment the wall of impossibility of the physical mind was crossed, Mother stumbled upon a world so utterly supple that it was bewildering—frightening, too (this is the other side of the difficulty, the one we were not expecting), just as if this terribly rigid matter were only the matter of our own fear. Mother had told me so for years, but I could not quite grasp what She said because, like all my human brothers, I am saddled with my own particular physical mind—until the day a little experience completely cleared up the "problem" (I should say "deflated the problem"). It concerned the beginning of a tumor in the neck. And Mother explained: *Probably a hair that curled under and the body covered it with a layer of skin and, out of habit, continued making skin over it: one layer, then another layer . . . It's an idiotic good will.* The whole cellular story was there, in those three words: idiotic GOOD will. *And that is how it is for almost all illnesses. The trick (there is a trick) is to tell the cells that that is not at all what is expected of them, that what is expected is not at all for them to gather together there, in a bundle, that such is not their duty—convince them. It's rather peculiar. It's the origin of habits. You see, they feel: This is what we have to do, this is what we have to do, this is* . . . [Mother turned her finger in a circle] *It was the same thing with me, but I told them. Only, one must be conscious of the movement.* [That is the whole point. We are not conscious of the movement, not clear. Everything is covered over by the mental hubbub, so of course we need microscopes and surgery—but it is simply a *movement,* a current.] *Then, very quietly, but very, very confidently, you say, the way you would to children: "No, this is not what your duty is. This is not your duty...." All chronic illnesses come from that. An accident may occur (something happens, an "accident"), then a kind of submissive and unconscious good will takes over and makes them repeat: we must repeat, we must repeat, we must ... And it stops only if a consciousness is in contact with them and makes them understand that, no, in this case, they mustn't repeat!* And Mother laughed heartily. *Oh, it's fascinating! But one must be very modest to do this work; one mustn't crave brilliant displays—very modest. And very quiet.*

Probably not everyone will know how to pierce through the layers and come into contact with the cells to "speak" to them. Actually, Mother did not expect to find that kind of a hero of the microscopic, for it takes "quiet" heroism; She did the work for all bodies. But what is of paramount importance in its simplicity is the *fact*, the phenomenon in itself, because it is a human, terrestrial fact. There are billions of absolutely identical little cells throughout the world—and it is a world of passive *good* will. Not a single law, not a single rule, not a single fatal and intrinsically defective molecule of DNA; just plastic matter spinning whatever we wish. We supply the impetus, and it goes on and on.... It latches on to anything and simply keeps on going. But this is quite extraordinary, if we think of it! Why could it not latch on to a vibration of joy and sun?... There is no obstacle, NONE. There is total freedom, a completely malleable world! Mendel's revolution in reverse. The physiological future unblocked—oh, how one breathed with Mother! She quashed all the myths. But, of course, this does not happen in three days (at least seven were required by the Almighty), and because we do not see the miracle happening instantly, we do not believe it—but the miracle IS ACCOMPLISHED. *The body is learning its lesson—all bodies, all bodies, all bodies. It is the change of authority. It's difficult, it's hard, it's painful. There is some damage, of course, but ... But something is happening—something is happening. There is de-*

*finitely something different in the world.... This was the work Sri Aurobindo had given me. Now I understand. And I see, I see now how his departure (Sri Aurobindo's) and his work—so immense, you know, and constant, in that subtle physical—how much that has helped. How He helped prepare things, change the structure of the physical. Truly, it is not as it was—it is NO LONGER as it was. And everything, every circumstance is as catastrophic as can be; absolutely everything is raging like a wild animal, but ... it's all over. The body knows it's over. It may take centuries, but it is over now. It's sure—it's sure and certain that that totally concrete and absolute realization one could have only when one left matter will be possible RIGHT HERE.*

It is the end of the "mental barbarism"[1] Sri Aurobindo spoke of. A barbarism from top to bottom, from religious summit to genetic code.

An open, fluid world where everything is possible.

A different little golden vibration is being spun in bodies as imperturbably, as quietly and irresistibly as our old cancers, which were but the cancer of the mind. Ever since that day in 1968 when Mother's body was left to itself, cut off from its memories, its mind, its old forces, the mind of the cells seized on the last thing it could to avoid being dissolved into nothingness and disintegrating for good (it was on the verge of disintegration, real death, for something is needed to keep the cells together, a vibration, a cluster of repetitive, habitual forces, otherwise it crumbles and everything crumbles); the cells seized on the Mantra, the Consciousness, the one great Current left in the midst of that general rout of all the little genetic coagulations. And they started repeating it twenty-four hours a day, day and night, without a second's respite, like a mule, as regularly and imperturbably as they used to repeat the old round of death.

When everything is gone, *that* remains.

The supramental vibration.
Matter's pure, true vibration.
The original fire that had lit the stars.
But it is an awesome fire.... "A vibration whose intensity is like that of a higher fire," said Mother. And which feels like love.

And with her never-forgotten humor, Mother added the following, which gives us at once the key to old matter spinning its little cancers and the key to new matter spinning ... something ... the secret of the future in the making: *There are instances when that power of repetition is extremely useful! I even think that this is what gives stability to the form, otherwise we would change form or appearance, or we would become liquid!*

It is the coagulant of forms.

Matter is only a vibration repeating itself.

Matter is neither hard nor opaque nor thick—it is nothing that our eyes see, nothing that our hands touch, nothing that our fabricated reactions feel. The quality of vibration is what causes more or less thickness, more or less opacity, more or less death—a lighter or heavier coagulation. Death is the scattering caused by loss of the customary coagulative vibration. Change the vibration, and matter changes. Density changes. And so does death.

Such is the secret of transformation.

The secret of the next body.

A new matter born from a new vibration.

Or maybe an eternal Vibration.

The only thing is to know whether the old body can bear the change of vibration without dying from it, transform itself without leaving the flesh.

A dangerous ... unknown.

SEVEN

# A Willing Automatism

To go through the second web is at once the most simple and most difficult thing there is. Maybe because it is, or ought to be, extremely simple in a certain way. Of such a radical simplicity, totally eluding the beings built on complication that we are, that it is rather dizzying. *Something fantastic ... which seems absolutely stupid,* said Mother after undergoing one last little "operation" in 1970. All the difficulties we thought of—inflexible laws, gravity and decay, coiling and uncoiling genes, cells wearing out and tissue wearing out, age limits and physiological impossibilities—were indeed the difficulties we *thought of.* It was *before.* They were the impossibilities and "laws" of the first web, matter as contrived by our heads and the laws of our physical mind. But when we reach the second web, the cellular substance free from its phantoms, the difficulty becomes the exact opposite of the old wall we have just broken through: the difficulty is that there are no more walls anywhere! It is that we float around in non-law, and everything is possible! It is rather frightening for a body. Nothing is solid anymore. It is at once marvelous and terrifying. Not a single mechanism to cling to, except the great Mechanism, and if we let go of it for one second, it is instant nothingness, vanishment into thin air. And where does that nothingness, or great "something," lead? We realize that the wide world we have built is a towering illusion of death and disease and gravity and electron microscope, while the other . . . rather eludes us. It is still nonexistent for a body. It has never been experienced by an earthly body before, and there is no reassuring memory of it. In a way, it is as if one body, one first earthly body, had to colonize the unknown—enter nothingness, which will become something by the very fact one enters it. But each step forward is a step into nothingness. At least Columbus entered a solid unknown. *The annulment of the person! An absolute annulment and incapacity. . . .* Indeed, what person could there be? A person needs memories to remain standing on an "I." And what memories could there be, since they were all memories from the old cage? What "capacities," since they were all the capacities of dying brilliantly? So every step is a free-fall. And an hour is made of many little seconds . . . scores of little "falling" seconds. Nobody will know what Mother went through. She kept asking for the time: "What time is it, what time is it? . . ." And people thought that She made too much of it, that She was getting a little incoherent. Sometimes they even rebuffed her. Until the day She stopped asking for the time. *Really, it's the consciousness of the cells that must change, you understand? . . . And there are no words to express that because it doesn't exist on the earth.* It means building another person. What can happen in such a body?

## Direct Functioning

Actually, a lot of fantastic and marvelous and infernal and incredible things happened, which She did not tell anybody because they would have believed her mad—I always think of Sri Aurobindo's silence. But it so happened that, in the past, I myself had gone through a sort of human no-man's-land, in a concentration camp, which was perhaps remotely similar to Mother's cellular no-man's-land, and which freed me once and for all of all human rationality. After that, one *had* to believe in a higher earth, or die. I did not die thanks to Mother, because in her I met someone who was seeking to build that higher earth, and I was willing to understand *anything* but the human prison, the distressing physiological concentration camp that ends up in a hole only to start again with another, "improved" concentration camp. So I totally understood her. For me, the impossibility was the old world. For me, the "horrible thing" was a living and tangible reality. And indeed, under the Pressure of that nothingness, or need—that old evolutionary lever—something else did emerge in Mother's cells. A new way of being in the world.

But first, She had to remain standing on her two feet. She had to make gestures, pronounce words in French or English; there were ten, twenty, fifty people waiting at her door and a few others watching. There was that invisible and silent pressure of all the little human consciousnesses that wanted this or that, expected this or that; and money to distribute, and that house to rent or not to rent, repair or not repair—hundreds and thousands of absurd details that make a grating, agitated, harried life. Her body was immersed in all that, surrounded by an innumerable, agitated little death. It did not understand any of it. That did make any sense to it. What does a check mean? What does the quarrel of this person or that person mean? A numbing pain pervaded everything. She bumped everywhere into a world of incomprehensible pain, an aggressiveness as devouring in its displays of love as it was in its displays of falsehood. With, sometimes, just a clear little vibration that wanted nothing, asked nothing, expected nothing—a miracle. Whenever something started to quiver somewhere, She stopped, turned inward toward that miraculous little flicker, smiled with her eyes closed. A breath of sweetness. And it all started again. Whenever I sat silently near her, groping my way in the new world, I could feel her stretching out in that immense, quiet sweetness, so overwhelmingly powerful, and if, for a split second, in a moment of weakness, a thought happened to come toward me, just at the periphery, at a distance, She immediately jumped, her whole body pulled itself together, and She opened her eyes: What time is it? A thought, even the most innocent one, was a shock in that. *Thought is the enemy,* She said. The enemy of the cells. It is something I witnessed time and again. For the cells, it is consuming—for thousands of years they have been under that consuming spell. A thought is the master that comes and says: I want you to do this; it's cold, you're going to get sick; it's hot, you're going to get sick; it's late, you're going to get tired. . . . It is the world of instantaneous pain, the return of the old slavery. And don't forget to take your Coramine, otherwise your heart will stop and . . . She fought and struggled with all that. She lived every second in death. No, She lived death. To "live" was death at each instant—at least to "live" as those around her conceived of it. One day, as I was talking with her about the progress of the experience and as She was explaining to me one functioning, then another, out of some old, absurd human habit, I asked what would happen in the future (always that obsession with the "future," when we are not even capable of living the present second), and since She liked me, She *tried* to know, closed her eyes and turned her look inward, but She abruptly came out of it, pulling at her shawl as if She were suffocating. She was perspiring, on the verge of fainting: *Look, you see what happens. Now, when I try to know, I am overcome by such a suffocating heat I feel I am going to die. That's how it is. Do you understand?* And that is exactly it: a world where one cannot "try to know," a world where one cannot try to do or even want to do anything. A world where nothing is controlled by the mind anymore, not even lifting a spoon and putting it in one's mouth.

One does not "know" and cannot do anything, but just when it is necessary, things are known, that is, they are done: to know is to do; to see is to have power—automatically. And if something is not supposed to be seen, one simply does not see it; if it is not supposed to be done, one naturally does not do it—one does not even know *how* to do it. In other words, all activities take place in the reverse order to the human practice in which we first need to "understand," to "know" and to "want" in order to be able to do something. But since our understanding is uncertain, our action is uncertain and our whole life is uncertain. While there, the action is infallible. It is immediate, to the second. A check does not mean anything, you see, but the pen will land on the check or it will not—there were certain papers that Mother simply could *not* sign. She had even forgotten how to sign! And two days or a month later, it turned out that those papers were a falsehood. Or else, She could not touch a flower or receive an offering of money from someone—which was laden with calculation and poison underneath. For the most microscopic or "insignificant" details of life, She "knew" or She did not know anymore, saw or did not see anymore, spoke or could not speak anymore. *For instance, if I am not "supposed" to say something, instead of forming the thought: "I mustn't say anything"—I can no longer speak! . . . And all sorts of things like that. It's a direct functioning.* Nothing goes through the mind anymore. The mind loses its usurped position of creator and driver. *The supramental action is decided by jumping over the mind. The mind is an immobile zone of transmission. It is completely quiet, peaceful, and is set in motion only upon receiving an order, an imperative order. When it receives the order, it does a specific thing, for a specific reason, a very specific action, and then . . . back to silence and stillness.* The mind resumes its true role of instrument, like the insect's antennae, the crab's claws or the nightingale's throat. *And that rehabili-*

*tates everything. It's simply the end of the quagmire that people have made out of it.*

## A Smile That Knows Everything

She went through the long apprenticeship of the direct functioning, which was perfected only after a long period of groping in the dark, after the 1968 turning point, when the whole mind was withdrawn, leaving only the mind of the cells. Actually, it was not something one "learned" the way one learns judo or swimming; it was a *way of being* that created the right, automatic functioning. Or perhaps a way of not being, of being nonexistent. A transparency that causes everything to flow naturally in one direction or another. The more "nonexistent" we are, the more total we are! The more "I" disappears, the more the totality of the universe is right before us. The more we know nothing, want nothing, decide nothing, the more total knowledge and encompassing vision we have, in every nook and cranny of the universe—we are in the center of everything because we are in *the* Center. Then we know and we do. It is the body that knows, the body that sees. There are innumerable little cells everywhere. Everything communicates, everything is right there. The great Consciousness flows, innumerably aware of the least movement of each detail of its body. *The slightest mental interference from the old movement spoils everything,* Mother remarked. *I mean, the old way of behaving with one's body: I want this and I want that and I want something else, and I want my body to do this, do that. . . . The moment that shows its face, everything stops, all progress stops. One has to be in a beatific state, then one notices the other functioning taking over. But it is such a subtle play! The slightest little thing is enough: just an ordinary movement, a movement of the ordinary functioning; when you slip back into that out of a sort of habit (it's tiny; these are not things that are easily perceived; it's very, very, very slight; it requires a great, great deal of attention), if that happens, everything stops. Then you have to wait. You have to wait until it eventually stops, that is, until it falls into contemplation—goes back over the whole course again. And then, when you've captured it again, when you can stay in it for a few seconds, sometimes a few minutes (when it's a few minutes, it's wonderful), everything is fine. . . . Then it goes awry again, and you must start all over again.*

In other words, thought is automatic static. The mental world is the world of static. And of course, since everything is scrambled and confused, we have to invent scores of complications to clear up the confusion or, rather, to confuse the confusion a little more. It is the world where nothing is known directly, where everything has to be "planned" and "organized." It is the great organization of confusion. While there, nothing needs to be "planned"—plan what? Each second of the universe is utterly new, utterly free, utterly organized within a *unique*, innumerable marvel—ONE—in which the flight of the bird over the Arctic is exactly connected to the little breeze that makes this leaf quiver before our eyes. It is all ONE movement, down to the second. We just have to live that second, that's all. Or, rather, *be* that formidable second in everything. And this, down to the cells of the body. *It's beginning to obey another law,* Mother remarked. *For instance, to know to the second what has to be done, what has to be said, what is going to happen—if there is the slightest attention or concentration to know, it doesn't work. But if you are like that, simply in a sort of inner stillness, then you know all the little details of life, just at the second you need to know them. What you have to say comes: Say this. And not as an order coming from outside—it comes, it's right there. What you have to say is there, what you have to answer is there; the person who comes in comes in, without advance warning. It's a sort of automatic thing that you do. In the mental world, you think about the thing before doing it (it may be very quick, but both movements are distinct), whereas here, it doesn't work that way.*

A *willing automatism*[1]—such is the way Sri Aurobindo defined the supramental life. Instead of the unconscious automatism of the animal and the atom, it is the *same* automatism in full light.

And the great universal rhythm pervades the least detail of life. The great rhythm at each second, in everything. *There was nothing but that ... something ... what shall I say? The English word "smooth" is what gives the closest feeling— smooth, even. Everything is done smoothly, absolutely everything without exception: bathing, brushing one's teeth, washing one's face, everything. ... There is no "big" or "small," "important" or "unimportant." And it's something so uniform in its multiplicity—no more conflicts or friction or difficulties or— something that goes on and on, with such a smooth movement, without resistance. I don't know. It is not an intensity of bliss, not at all. That too is so even, so smooth, though not uniform— it's innumerable. And EVERYTHING is like that, within one and the same ... rhythm ("rhythm" is too violent a word). It isn't uniformity, but it's something that is so even and that gives such a feeling of softness, you know, coupled with a formidable power in the least thing.... No more memories, no more habits. Things are no longer done because one has learned to do them; they are done spontaneously, by the Consciousness. It's not: "Ah, I must go there"—every minute, you are where you're supposed to be, and when you arrive where you were supposed to go, it's: "Ah, this is it."*

This is *it* at every second.
We *are* at every second.
Or we are born, perhaps.

It is the world "without succession," without before or after, without fatal consequences—nothing is fatal! It is our head that is fatal and extends into the future its dark and morbid little ruminations, perpetuating disease, perpetuating death, perpetuating everything. *The Consciousness is constantly in action, not as a continuation of what took place before, but as a result of what it perceives AT EACH INSTANT. It is the Consciousness that CONSTANTLY sees what needs to be done. It is the Consciousness that follows at every second—it follows its own movement! And that makes everything possible! That's precisely what makes miracles or turnabouts possible—it makes everything possible. It's exactly the opposite of human creation.*

Nothing is closed anymore.

Each second is total, pure.

We have confined everything to our head, even time and space and chromosomes, but that is not how the world is; that is a caricature of the world, a scientific and mathematical hell. An illusion of the world. A "miracle" simply happens when our logic is punctured. Then there is an instantaneous miracle. The scientists are the last sorcerers. *We try a number of combinations . . . to end up building our own cage, then, all of a sudden, a breath—a warm, luminous, relaxed, comfortable, golden breath—why, it's obvious, that's it; I'll be naturally CARRIED wherever I have to go; what's the sense of all that complication! . . . You see, I am here in the midst of all these circumstances, these complications, these people, these—everything is in such a turmoil—yet, behind, there is a sort of . . . It isn't just a Force; it's a CONSCIOUSNESS-Force—it's a Consciousness—and it's like . . . a smile. A smile. . . . A smile that knows everything. That's it, you see.*

The smile of the next world.

*I am convinced that this is the transition from this life to that Life. When we are entirely on that side, well, we'll cease speculating, trying to "explain," trying to deduce, conclude, organize. All that will be over. If we knew how to . . . be—BE—simply be.*

EIGHT

# The Exit from the Second Web or The New Body

But how could this new way of being create a new type of matter?
The body moves in another way, knows in another way, acts in another way, but it is opaque in the old way, digests like everybody else, breathes like everybody else, and even if its heart obeys other laws (otherwise it would already have died a hundred times), it is nevertheless a heart pumping blood into veins. Even if the wear and tear has ceased, it is merely a reprieve amidst death. It is not the next species; it is the old species improved. Geneticists may be under the impression that, by moving molecules around or changing their order, they will, with luck, produce another being. But if such is their hope, they are mistaken. They will produce monsters, caricatures or perhaps, if they are lucky, improved superbrains. But these will just be different versions of the same thing, variations of the same basic substance. We can put together the molecules of Napoleon, Shakespeare and Dante—it would be interesting—but the result will be human all the same, and perhaps worse. What follows man is perhaps (certainly) produced out of man, but it uses an element that has nothing to do with man. A new

element that makes the whole difference—in what gene or molecule will they find that new element? We always think of the evolution of species as a continuous chain, but those are the species of one and the same animal Species. The next species does not belong to the animal kingdom. It is not another variation, but something else. A new evolutionary saltus, like that between vegetable and mineral or between animal and vegetable. A next kingdom. A new matter arising from the same eternal matter. Can we imagine a being made out of a matter as different from human matter as the matter of granite can be from that of a rose or a dragonfly, something that is neither vegetable nor mineral nor animal? That is impossible, we will say; it does not belong to the realm of existing combinations, or else they are bodies of fantasy, celestial apparitions and the whole esoteric almanac. But here we are concerned with matter, not with miracles, unless they be matter's miracles. *The physical mind,* said Sri Aurobindo, *always comes in with its fixed line of the present and "No farther" and when the fixed line of the present is unfixed and overpassed, it again erects a new line and cries "No farther." If an "elemental" who had attained to the physical mind had been present at the different stages of the earth-history he would have argued like that. When only matter was there and there was no life, if told that there would soon be life on earth embodied in matter, he would have cried out, "What is that? It is impossible, it cannot be done. Life is possible only in a subtle body. It has never been and will never be embodied in gross matter. What, this mass of electrons, gases, chemical elements, this heap of mud and water and stones and inert metals, how are you going to get life in that? Will the metal walk? Can the stone live?"*[1]

And now that only "life" exists, and no "overlife," are these DNA helices, these pulmonary alveoli and exquisite gray cells ever going to live in anything other than the good Lord's good (and slightly polluted) air, in a palpable and at last reasonable matter? It is impossible—we have to soar to heaven.

But how is that other matter created? What does it look like? It is not going to fall ready-made from the sky, is it?

## A Way of Being

We are approaching the most mysterious part of Mother's great Forest, and yet we feel it must be so simple. She did not know the way. She simply walked. She said: I saw this, I saw that, I had that experience.... And it seemed to come from anywhere and to lead anywhere. It was a *fact*—yes, but a fact of *what*? There were thousands of facts ... of something one did not know how to name or define or tie together. When a baby grows up, there is that "garden"—rocks, weeds, lots of unnameable and unnamed things that feel, live and happen—and then what? It is only much later that it makes a "garden." And sometimes I complained to Mother, even in 1972, one year before the end of the "garden": "I don't quite understand what course we are following," I said. *Well, I personally don't understand it at all! I just ... [and here She opened her hands in a gesture of abandon]. It isn't easy.* So if the reader thinks that I am trying to "make a point," he is mistaken. I would only like to understand that garden. I do not even know where it is going or what it is. It may be the new world, but it is nevertheless quite strange.

One day in 1970, one morning, Mother simply remarked, *I have the feeling something is trying to develop in the body, I mean a way of being of the cells that would be the beginning of a new body. When it comes, it produces a strange sensation. A strange sensation. The body itself feels it is going to die—something ... it doesn't know what it is. Only a state of very intense faith makes it possible to withstand. As if one thing were being changed into another. As if what is now were trying to change into something else. And that's really . . . hard. Something entirely new.* There lies the whole mystery. A way of being of the cells that would be the beginning of a new body. . . . How can a way of being form a new body?

Evidently, there is a way of being that makes a lizard, a way of being that makes a flower, a man. What is it that makes a lizard after all, in the beginning? Always that "beginning," that "very first time when." We come after the fact and say: It's quite simple; it coils up into little molecules of protein according to such and such a pattern, and if the pattern is disturbed, it no longer makes a lizard; it makes—what? But what made it coil that way in the first place, made it want to coil that way and not any other? At that point we conveniently invoke "Nature" or "evolution." It is simple. But where is that lady, or gentleman, the something that willed in the beginning—unless we think nothing willed anything, or the good Lord, perhaps? We always have the impression that scientists conceal the good Lord under their Greco-Latin molecules. So something willed—things must be *willed* in order to be done, there is no question, or they must be envisioned in order to be done, or they must *be* . . . something. It is the parents who grow in the baby. But in an amorphous globule of gelatin, there has to be "something" that grows (calling it "life" simply shifts the mystery under another label), something that "tends toward"—a something, that means a being, a self-consciousness, even it does not resemble our superb cogitations, a way of vibrating or tending that uses whatever means it has to spin or secrete its own existence. If we invoke the sun and amino acids and the bombardment of particles, under so many degrees Fahrenheit, it only amounts to saying that the prison has produced the prisoner. It is a science of prisoners. Even those amino acids willed to be something. They have their own way of vibrating—their way of being. This is not philosophy, and I could not care less about philosophy. I do need to be a Marxist or even a spiritualist. I am merely trying to understand the phenomenon. The phenomenon is that of a body or, more precisely, that of the cells of a certain body that have lost their habit of spinning within the grooves determined by the physical mind, but that have not lost their habit of being—to be means to beat, to vibrate, to strive... perhaps to will, but without quite knowing what they will or where they are going. In short, cells that would be at the very-first-time-ever stage, but after having described the entire human course. What are they going to spin? What substance? What will happen? A being necessarily produces a body for itself; a way of being or vibrating must coagulate matter or substance in its own way—whatever is right there, as it were. And what is right there?

If we knew what is *there,* the pure thing to spin outside of the old patterns, we would be very close to touching the secret of the world. We have perhaps followed this enormous spiraling journey through evolution in order to reach the moment when, individually formed and individually conscious, we can touch the secret of the beginning, the key of the beginning, the energy or matter or being of the beginning—what is right there, pure, without all the evolutionary crutches that have supported us thus far. Then we will move from the science of the crutches to the science of being, from the science of our old, successive prisons to the science of freedom. For once, that would be a fascinating science.

## The Exit from the Second Web

Here, we have nothing to hold on to but "facts" (even though we still do know the facts of what), that is, the experience of those pure cells as Mother was falteringly trying to tell it. And there are so many experiences, sometimes almost contradictory, though perhaps no more than the rocks and weeds of the garden contradict the pond—we simply do not know their connection very well because we do not know the "garden."

The phenomenon that seems to recur most often at the beginning is a state of cellular fluidity (or fluidity of the cellular consciousness). In other words, the cells' principle of amalgamation—the customary vibration that holds them together, that makes an "I" in the form of man or lizard, the something that repeats itself—seems to dissolve. *It's as if there were a dilation—a dilation—and like something that is trying to melt. It's a very, very strong sensation. And that produces an extraordinarily powerful vibration through the cells, something quite out of proportion with the human body—formidable!* That vibration is just what our mystery is. We say it is the supramental vibration, but we certainly would like to know what that supramental is all about. *And when that happens and I look around, I see people melting (not many, very few), while others are frightened out of their wits, and get up and run away!* ... Indeed, when one felt "that" going through one, it was rather ... formidable, but formidable probably because the human body was not at all accustomed to it—a vibration so foreign to its very substance that it felt almost like a threat. *I just have to stop all [external] activity, only for two or three seconds, at most a minute or two, for the body to feel it is floating, floating effortlessly, floating. And I see an immensity, a sort of ocean of that luminous, golden, powerful, vibrant Consciousness. And the body is floating in it.* ... And we wonder if this is not what primordial matter is. The "something" that each one and each species has spun, channeled, concretized or fossilized in its own way: the world's primary substance, the vibration from which all other vibrations stem, or of which all other vibrations are a reduction, a deformation or formation in the size of a flagellate or a lizard. What we call "matter" is a certain way of spinning or confining that, and there are all possible degrees of matter.

So what will happen to a body that lets this pass through it unhindered? Is it going to dissolve in it or "confine" it in a different way? What can be the new principle of amalgamation, assuming it is at all possible?

The beginnings of the experience are very "disquieting" for the body. That "dilation" feels very much like dissolution. There is no longer any bodily "I" diligently repeating its little coagulative vibration, the ceaseless, invisible trepidation that holds everything together. *How do you keep a form without an ego? This is the problem. And it's precisely the fascinating experiment that is taking place right now. At times, you feel that everything, but everything is dissolving, falling to pieces. And I saw what happened: at the beginning, the physical consciousness was not enlightened enough and so, whenever those inner preparatory experiences would occur, it felt that, oh, this must be a sign announcing death! Then gradually came the knowledge that that wasn't the case at all, that it was only an inner preparation so I would be apt. And I very distinctly saw that, quite on the contrary, if that very special plasticity, that extraordinary suppleness is realized, it will evidently mean the abolition of the necessity to die. In other words, an unknown state. A state not physically realized, one could say.* The current is no longer confined by anything, so there is obviously no longer any reason to die, for it is the confinement or hardening of the current that creates the necessity of death. But how can one remain standing on one's feet in that "plasticity"? It would be Mother's problem for months and months. And to some degree it would continue to be a problem until something new began to develop in the body, a new sort of "supple solidity," as She called it. What was difficult was the transition. *Spontaneously, when left to its old habits and ways of being, the body is in great difficulties—it results in an inner organization that looks very much like disorder. It's difficult. You see, problems constantly arise, for everything—everything. There is not one bodily activity that is not called into question by that [the fluidity]. Eating is becoming a problem, sleeping is becoming a problem, speaking is becoming a problem—everything is becoming a problem. . . . The process is no longer the old process; things are no longer the way they were, but "the way they are" has not yet become a habit, a spontaneous habit. In other words, it isn't natural; it requires the consciousness to keep a watch on everything, even on swallowing food. . . . Oh, it's so difficult! A way of being that is unnatural according to the old nature—but which is evidently the new way of being—has to be maintained by a kind of conscious concentration. And that's almost a Herculean labor.*

That new way of being is first of all a way of not being, because "being" means to be all the old habits of being. It is something that naturally has no longer any particular center, no longer any prison. It spreads out everywhere. *The body feels the forces coming, but . . . it doesn't even feel them passing through. They just pass through without . . . Through what, I don't know. It's very intangible. And if there is the beginning of self-awareness or self-anything, it becomes extremely unpleasant—a discomfort, an inexpressible discomfort. This has no limits, you see [Mother pointed to her own body], that's what is so curious. . . . For instance, there is a phenomenon (one among many), a peculiar phenomenon: I don't have the feeling that "I" eat, you know, and so I am not conscious of putting things into my mouth and of needing to swallow them and so on. . . . No, it's something that's both in me AND IN THE FOOD. It isn't something that comes from the outside; it's like something . . . [Mother drew a sort of movement of forces in the air], something developing, something free to develop. Then everything is . . . fine. But the minute I become conscious with the old consciousness, which means eating, tasting the food, putting it into my mouth—it becomes difficult! I have a world of trouble not to swallow the wrong way. . . . There is no longer any sensation of the "something through which," the something through which the Divine passes—it's gone. It's immobile and nonexistent. It has no self-consciousness. It is only conscious of . . . the divine Action, and that's all. Then everything is all right. But the moment there is the slightest sensation of something "passing through," the discomfort comes. Listen, I could say (it sounds like literature!) that in one case, in the state where the body is no longer aware of itself, where only the awareness of the Divine is left, there is a sense of immortality and Eternity, while if there is the least sensation of "something in which" the Divine manifests, it immediately turns into the sense of death—one immediately becomes mortal again.*

One falls back into the prison.

One immediately starts spinning death.

*It's become a very acute condition,* She added. And it was going to become more and more acute. *For the least thing I can't swallow anymore, or I can't even breathe anymore. . . . It gives the feeling of a life that's about to depend on something other than the usual conditions. But the new conditions are not yet there; the body isn't accustomed, and that switching from one state to the other is what creates a perpetual difficulty.* It was a kind of constant paradox in everything, a radical and dangerous contradiction every minute of the day, actually between the state of life and the state of death, the new state and the old state. *It's something incredible: either the true consciousness or the sense of impending and general danger. In everything, you know: eating is a danger, taking a bath is a danger. . . . It's as if the body were shown in all sorts of circumstances—countless circumstances—how one goes toward death and how one goes toward life. In everything, absolutely everything, in every part of the body, every organ, one activity after another—it's impossible to describe. . . . And it's strange, the moment the attitude slackens in the least, for example, a second of forgetfulness (what could be called forgetfulness, that is, when the old habit from before, the old earthly habit returns), the body immedi-*

*ately feels on the verge of dissolution. And that moment, oh, can last two or three seconds, like that!—a feeling that everything is going to dissolve.*

And sometimes, Mother no longer knew at all where She was, on this side or the other, going toward life or toward death, disintegration or something else, and She would let out a cry: *You know, I feel I am plunged in a world that I don't know, struggling with laws that I don't know, in order to work out a change that I don't know either—what's the nature of that change?* And I tried telling Mother what I deeply felt as a sort of intimate and obvious truth: "Yes, but, Mother, I have a very strong feeling that, through that darkness and ignorance of the 'laws,' you are KNOWINGLY taken to the point where the solution will be found." *You're right. You're right, and in a way I could say that I think along those lines (I don't "think," but anyhow . . . ). There is a sense of that. But . . . there's everything in between.* And She laughed, making fun of me: *Go ahead, continue thinking that way!* And I protested, so sure was I of the deep logic of that hell. I could almost see the other side of that no-man's-land: "It simply is impossible not to succeed!" *Why?* And it came bursting out of my heart, as if the earth's entire hope were there, in that impossible transition: "Because . . . because you are the body of the world! Because it's really the Hope." *Isn't that poetry?* This was so typical of Mother, this "poetry" She rejected, of which She was perhaps ashamed, for, after all, She was *making* the "poetry." "Of course not, Mother! It's not poetry: that's how it is. It is quite clear that the outside world is getting more and more hellish." *That's quite true.* "Well, that's what is in your body."

It was that indeed. It was the passage to another world, another kingdom, a true earth, a free earth, on the other side of all cages.

The exit from the second web.

The physiological and genetic cage.

It was in 1970.

## The Key to the New Body

In fact, the solution She sought was there, at hand, unknown to all. It was *in* the difficulty itself. The transition from the mineral to the plant's animated (but still static) life must have meant a prodigious scattering of the cage of stone, and that from plant life to the animal's movement, another staggering dispersal. And the next one? What is neither life's movement nor the rock's static immobility? The contradiction or paradox grew ever more acute in that animal life which was dying in order to give birth to God knows what. *There are moments when the body feels such a great force that it could do . . . anything—a force of another nature, but far more powerful than before. There are moments when it can't even stand up, and for a reason that is not . . . physical: It is no longer governed by the same laws that hold people up, so? . . .* And this: *A strange experience. It's a strange experience. The body feels it's no longer part of the old way of being, but it knows it still isn't in the new way. It is no longer mortal, and it is not yet immortal. It's quite strange. Very strange. And sometimes, I go from the most dreadful discomfort to . . . a marvel. I have to be constantly concentrated, concentrated to be able to do things* [and one could see Mother trying to pull toward "herself" something that was spread out everywhere]. *Sometimes, not a single word in my head, nothing. Sometimes, I see and know what's happening everywhere. . . . I have to be careful when I am with people, otherwise they might think I've become mad. It's truly strange. It's like total powerlessness and extraordinary power at the same time. But sometimes I can't even eat!* And I asked Mother, "If only it were possible to know what exactly causes the switch from one side to the other? . . . " *Yes! There's obviously an at-*

*tempt to make the body know. And sometimes it suddenly finds itself . . . outside of all habits, all actions and reactions, all consequences, etc., and that's . . . a marvel. Then it goes away. It's so new for the material consciousness that each time you feel as if . . . on the edge of an abyss. For a minute, the consciousness is thrown into panic. From the very beginning and continually, there has been a sort of deep-rooted common sense in me that rejects any notion of fantasy and says: I don't want to fantasize about this, I don't want to fantasize about that. . . .* Past ninety, Mother still appreciated Mathilde's virtues. *And so the consciousness accepts things only when they are altogether concrete. It's all too easy, you know, to start embellishing things and—out of the question. It must be entirely practical and concrete. Therefore, I am sure this is not the result of mystical dreaming in me, absolutely not. This body had nothing mystical in it, thank God!* And this is where I said to Mother, "If all of a sudden, through accelerated evolution, a caterpillar were given the eyes of man . . ." *Yes!* She exclaimed. "It would be frightening." *Yes, that's it.*

And that was exactly it.

Yet, as if implicitly, something else seemed to want to be born or show through. But when one is in it, one simply does not know—one does not see oneself in the process of being. One does see what is going away; but what is coming is so new it is invisible. The eyes have to become accustomed for it to become visible. Perhaps we need eyes adapted to the new world—what does a man mean to a dragonfly? Does he even exist? There is not any pollen on him, or any pretty sparkling of fresh water . Our eyes are terribly functional, but what happens when the function ceases, even our esthetic function? *The consciousness of the body is slowly being altered in such a way that all its prior life seems alien to it. It just seems to be somebody else's consciousness, somebody else's life. Its "position" in the world is altered, if you will. It's as if the past did not exist. You are as if totally thrust ahead, you know, like that, and there's nothing behind. A strange sensation. The strange sensation of something that's beginning, not at all of something that's ending. It's really a strange sensation—something that's beginning. Along with all the sense of the unknown, of the unexpected. ... Strange. I have constantly the impression that things are new, that my relation to them is new. And the body, too, has the impression of a new way of feeling things, a new way of reacting. ... It's very strange.* And there was always that "inexpressible discomfort" flaring up at each step, each movement, perhaps a hundred times a day, as a sort of indication of the back-and-forth movement between the cage and something else, the marvelous opening followed by falling back—perhaps falling back to be able to jump even farther? We understand nothing of our difficulties; they are always our springboard! *Generally, when that discomfort comes, the body immediately surrenders—it surrenders as if saying: If this is death, well, let Your Will be done. Total surrender, you know. Then, sometimes, depending on how ... successful the surrender is, I don't know, it sometimes produces a clarity, a comprehension, a sense that EVERYTHING IS OBVIOUS.* And we wonder if that other, emerging life, that other kingdom which is no longer mineral, vegetable or animal, that other inconceivable movement of being, would not be innumerable life. A life no longer confined within a shell or an integument, or any skin of rock, tree or man, but something running innumerably, being innumerably—a sense that everything is obvious. A state of obviousness. For us, nothing is obvious. We always have to run after things, and once they are caught, we still have to "look" at them. Here, we are simply in them—they are obvious. But where is the body capable of such a feat? For it is a question of doing this with a *body,* not just a roaming consciousness. *A truly remarkable state. But it doesn't last. The slightest thing will disturb it. I know, the body feels that if it could TOTALLY surrender, no longer have a separate existence, no longer exert a personal effort, no longer have a personal will . . . to the degree that that is possible, everything is fine.*

Surrender, indeed, but how to become the new while still being hooked up to even the best physiological processes of the old?

Total surrender, for the body, means accepting death. Accepting it physiologically. Accepting it mentally is fine, but what do you do when you start suffocating?

Yet this is where we begin to touch the key or lever. During that inexpressible discomfort into which the body was thrown, that sort of return to death (a real suffocation—a body suffocates when it approaches death), something happened, always the same thing, very simple, automatic, repeated thousands and millions of times, because a day and a night are made of many seconds (86,400): *An intense aspiration in the body.* It is the simplest phenomenon there is. There is nothing left, so one has to cling to something. It is the very movement of the first respiration of the world, which must have been an aspiration. A suffocation of non-being that wants to be. It is the deep throb of life, of all life. The first spinning of something around a nucleus. The secret prayer in things. The hidden name of beings. It is there, deep inside, but generally it has to cross layers upon accumulated layers, and it spins nothing but a habit. But that first cry of matter, that need to be, or perhaps to love, is at the origin of each species, each thing. It grows with each species, each step of the long journey and, sometimes, stripped of everything, perhaps broken, it bursts forth from the very depths of our body. A burning. An intense heat that feels like love, something that is very still but feels like a compressed, compact, almost overwhelming power—an overmuch of something that is both quite pleasant and quite unbearable, as if we were on the verge of exploding. It happens *in* the body and seems to arise from everywhere at once. It has nothing to do with feelings; it is like a tidal wave of flame. And strangely, it loves. It is as if we were suddenly filled with the most profound anguish, the most intolerable void, almost a pain, as well as with a fullness of something that seems like *ourselves* for the first time ever. Everything else falls to dust, but that *is,* sovereignly. It is even all there is. Seconds of death that are like a sovereign and inexpressible life. Twenty years later they still shine like pure gold, as if they were eternal moments in the body. One day, I asked Mother what that sort of intense aspiration was which sometimes arose in the body. *I think that what we call intense aspiration must be the supramental vibration.* This is what is taking place inside bodies—all bodies—that original vibration which has spun life into countless forms, which has become encrusted, hard, simplistic; but a little breaking and a little scratching of the habit, and it comes to the surface. It comes instantaneously. It is the being of what moves, sleeps, eats and kills and forgets, so much forgets what it is. *In each thing, there is that luminous, golden, imperative Vibration—which is necessarily all-powerful.*

Then we have the key to the new body, because it is the key to all bodies ever made on this planet. It is the primary matter of the world, true matter.

A sort of very particular wave.

## Primary Matter

We can fairly easily visualize that "powdering of warm gold" which was Mother's first direct supramental experience in 1958, twelve years earlier, when She went all the way down into the "Inconscient," in other words (we understand it now), when She pierced through the crust of the physical mind—the rock of the Rishis. *Like a powdering of atoms vibrating with extreme intensity*, She said. Indeed, our atoms themselves could well be the first coating or first product spun by that fundamental vibration. *An immensity made of countless imperceptible points. A multitude of tiny golden points, nothing but points. It seemed as if they touched my eyes, my face—and there were such a power and warmth in them! It was fantastic!* Today, I seem better to understand the course traveled during those twelve years: the slow preparation for crossing the web of the physical mind; the widening and universalization to withstand the "boiling porridge" of the supramental powdering without disintegrating—*that Movement which exceeds the force or power necessary to hold the cells together,* She noted in 1963—the slow clarification of the cells from their opaque periphery to get to the nucleus, the primary vibration free from all the layers of the old spinnings—that "vibrating immensity" in which at times She felt about to dissolve. It was like a return to the material origin of the world. And the experience is identical each time, year after year (as if there were small touches of experience to accustom the body progressively), and still in 1969: *There is an intense aspiration and at certain moments—when it produces a sort of expansion—I don't know what happens, but something happens in the cells, like . . . a state, a state of intense vibration in which you have at the same time a sense of absolute power, even in this, in this old thing [Mother pointed to her body], a luminous and STATIC omnipotence, in other words, a sense of eternity in the cells. . . .* This is the strange contradiction of that Supramental, which seems to combine or integrate immobility and extreme swiftness, perhaps so extreme in its vibration that it is perceived as immobile, not unlike the immobility of the rock in the midst of its lightning-swift subatomic movement. And we recall Sri Aurobindo:

*A fiery stillness wakes the slumbering cells*[2]

## THE EXIT FROM THE SECOND WEB   71

*Something entirely, ENTIRELY new for the body.* Yet this was in 1969. What could be so new in that experience which seems so identical to many others already mentioned? Maybe it was not the experience itself that was new, but the level at which it took place. As if, as the years passed and more layers were crossed, the experience—the same eternal experience—became purer, more material, more corporeal, touching the very heart of the cellular substance. Once we touch bottom, we speak of "Supramental," but it really acts everywhere, at all levels, through mental force on the spiritual summits as well as through feelings, instincts, formations, deformations or depravities—through everything. It is the one and only driver that runs through complications or obstructions, through countless, diverse types of spinning—and finally the experience reaches purity in a little cell. *It's a state that seems to be completely immobile. . . . I don't know what it is. It isn't immobility, it isn't eternity. . . . I don't know, but it is something—it's something which is . . . yes, it's Power, Light, and really Love. Something . . . To the point that, when you emerge from that state, you really wonder if you still have the same form!* And She laughed.

Of course! It is the original coagulating agent of all forms. Something that combines the apparent immobility of the mineral kingdom with the increasingly accelerated movement of the vegetable, animal, and mental kingdoms—a new acceleration in apparent immobility? But what *is* that matter? . . . We talk of "primary matter," which is all very fine, but what does it look like? And how is it handled? Matter is not wind, although there is gaseous matter from which the stars have been formed. Scientists even tell us that matter is mostly empty, its minuscule nuclei being distantly surrounded by their coats of electrons. That is what they have seen at the end of their microscope. And what did Mother see at the end of her direct microscope, "on site," so to say? Strangely, her repeated experience—repeated dozens of times—links up with Théon's at the beginning of the century, when he spoke of a "matter denser than physical matter, but with features that physical matter does not have, like elasticity, for example." And what is most remarkable is that a note describing an experience She had had around 1906 was found in Mother's papers after She left, a note She had completely forgotten about, which was written in pencil on the back of invoices for paintings of the "Atelier Édouard Morisset" (her father-in-law, the one who painted the portraits of the little Egyptian princesses) and in which She recalled the following: *Suddenly, I felt I was drawn into a breathtaking fall* ... [that "fall" looks very much like the sudden crossing of all the layers of consciousness or memories accumulated by the evolution of bodies], *a fall that seemed more and more breathtaking, finally to stop, as if stuck, in a place that I could not quite make out at first, where I felt a very strange sensation, one I had never really felt before: I felt I was in an environment denser than the earth itself, an environment that seemed as dense as a diamond, but which was elastic, surrounding me so closely that I could clearly feel the contact of that substance on my entire body and particularly on my face, arms and hands (the parts that were bare). The feeling was not unpleasant, but it was so new that it surprised me. Then the thought came to me that I might as well rest there a little to become accustomed to that environment, which I did, and after a while I found myself at ease and saw that that substance was slightly self-luminous, multicolored, and with molecules of different densities. There was also some self-luminous gold, but very different from essential gold, a little different in color but mostly because of the difference of density; that luminous gold was not transparent. Gradually, I saw a large sphere of that substance form AROUND me, and that sphere had every color.*

That substance formed *around* her. This is most interesting. And here we find again our "speckling of multicolored, iridescent light." But what caused it to form or agglutinate around Mother's body? Here is a question She spent perhaps sixty years solving or, rather, living. What She remotely saw, in a "vision," at the end of a "breathtaking fall," She spent sixty years touching directly in her body, with her eyes wide open, after having gone through all the layers, which are for us like layers of sleep, layers of false matter, or dead matter, we could say, stacked on top of one another by the evolution of bodies—all the residue of the mineral, vegetable and animal life which made up and still makes a body. In 1961, when for the first time She spoke to me (still rather "remotely") about that "other matter" or "new substance," having completely forgotten about the experience of 1906, She said, *That New Creation is something denser, more compact than the physical. One always tends to think of something more ethereal, but that isn't the case at all! The impression I have about that atmosphere is that it's something more compact—more compact and at the same time without weight and thickness. But solid! Oh, with such a cohesiveness to it, such a MASS, and yet ... I don't know, it's completely different than you'd expect. Completely. You can't imagine what it's like.... Something that is compact and without division. I mean, you feel you are on the wrong track: ordinarily, when you look for the "Supramental," you always tend to look for it above. But that's wrong! That's wrong. And then you expect a kind of increased subtlety, etherealness—but that is wrong.*

There is clearly a logic and a continuity in that experience spanning sixty years.

In 1967 again, She said, *Here, it was like molten gold—molten and luminous. It was very thick. And there was a power, a WEIGHT to it, you know, quite amazing.*

And now that her body was directly in contact with that primary substance, what was going to happen? What we call "matter" is obviously something "stiffened," as Mother said, a fixed and stereotyped movement, a force imprisoned right down to the atom, and it is because it is all stiffened, fixed and hard that we can grasp it—we touch the prison. The prison is "matter." In short, we can only touch that which vibrates slowly enough; for the slower it is, the more opaque it is. We can only grasp what is opaque. There is a whole gamut of light too rapid for us to grasp, of sounds too "high" for us to hear. A whole "spectrum" of existence escapes us altogether. And beyond the perception of our instruments lies a whole range of primary matter without opacity, too rapid for our senses—a matter that is linked to the very movement of consciousness. But we immediately dig a sort of supernatural chasm between palpable and verifiable matter and that imponderable quantity we call "consciousness." This is our fundamental error. "Will the metal walk?" asked Sri Aurobindo's imaginary being at the beginning of the earth, when life did not yet exist—only supernatural matter or, rather, supernatural and disincarnate consciousness, can possibly "walk" on that crust of minerals. We may fall into the same error today with all our scientific apparatus: Only supernatural matter or some disincarnate consciousness could possibly walk on the face of this good earth with something that is not a derivative of the mineral, vegetable and animal kingdoms—i.e., all the successive hardenings in evolution. We are only in touch with what is hardened. Everything eludes us because we have not grasped matter's central secret: Matter = Consciousness. Nor have we grasped evolution's central secret: Evolution = development of Consciousness. Because it does not exactly fit any of our catalogued, fossilized routines, we conclude that consciousness has nothing to do with matter—but this may be as great a superstition as saying that matter has nothing to do with energy. Or do we perhaps believe consciousness to be a product of all our little improved prisons? A kind of higher secretion? But it is the world's very matter. And the whole evolutionary experiment now in progress, the evolutionary challenge being flung more or less brutally in the face of all those little walking metals, is for this end product to rediscover what set him in motion in the first place and to build, with his conscious cells, a body of conscious matter, which will perhaps stand upright by using other laws than gravity, but which will walk on the face of this good earth as securely and solidly—perhaps more solidly—and no more "supernaturally" than today's little thinking metals. The secret of the beginning is at the end. Matter is not betrayed by consciousness; it does not become "subtle," nor does it vanish into a cosmic dream ("It was very thick," said Mother)—it enters a new acceleration and a new kingdom. The kingdom of conscious matter.

Matter without prison.

## The Mystery of the Unknown

But a *form* means a boundary or a structure. How can a non-supernatural body with a recognizable form not be a prison? A body that does not fall from the sky, for God's sake; we are in a logical and sensible evolution, even if it does not conform to our present sense or our rigid logic!

The experience is simple.

In that sort of dissolution of the form which seemed to thrust the cells into nothingness, or into an "ocean of vibrating consciousness" where nothing of what they had slowly, painfully built through the millennia of evolution was left standing; in that sudden suffocation—perhaps that negation of all that had made them throb, hope, live in bodies upon bodies—they were seized by an intense aspiration: to be again and again, to be forever! This was why they had been created. Death was a frightful negation, even a death in the light. Those cells were made of matter. They were calling out for matter's truth, matter's life. They clung to the Mantra, the little golden vibration inside; they kept repeating and repeating their prayer of being, their love of being; they spun and spun that one golden, dense substance, as the plant spins sunlight, as the butterfly flies straight to the pollen, as simply and blindly as that—it was a question of life or death. There were no more memories to spin, no more "I" as in all the other bodies; there were only those thousands of pure little pulsations in the depths of the cells, which expanded, gorged themselves with that only air left. A way of being on the border of death, a way of calling out and praying as at the dawn of the world, when there was still nothing but that pure little vibration striving to be forever, to love forever. Something very simple, so simple that all the words we put on it seem silly and pompous. "Isn't that poetry?" She asked. Indeed, it was truly the original poetry—it created. It created a body. A new body, slowly, day after day, year after year, as the shellfish spins its calcium. A spinning of dense, iridescent, at times golden substance around that nucleus of prayer or love in the heart of each cell; something that wrapped itself around that form, modeled itself on it, perhaps slowly filled it. Or perhaps absorbed it or gradually changed it into something else. *A strange sensation. As if one thing were being changed into another. As if what is now were trying to change into something else. And it's really . . . hard.*

She herself did not know for sure. She herself did not fully understand. She no longer had a mind watching itself being on stage. She was only thousands of conscious little cells calling out, calling out day and night, repeating and repeating the Mantra, *like a hymn . . . an incantation, you know, a call, an incantation to the supreme Power.* And if they stopped calling for a second, it was instantaneous dissolution, the "abyss." That's all. Mother was like a prayer of matter. She was a body, a certain body of old matter in the process of going into an unknown, other state, which felt it was dying at each second in order to emerge at each second into something else it did not know, did not understand, did not even recognize as another body. *There comes a moment when—the word "anguish" is far, far too strong—but it's the feeling of reaching the point of . . . the unknown. Exactly that. The unknown, the . . . something. And it's a very, very odd feeling. Almost constantly, the body has really the (at least very odd) impression of being—of no longer being this and not yet being That. Indescribable. But strangely, there's no fear whatever, no sharp sensation, no sharp sensation at all; there's something. . . . Well, it could be described as a new sort of vibration. It is so new that—you can't use the word "anguish," but it's the unknown. The mystery of the unknown. And none of this is mental, you know, it's just how the vibration is felt. And that's becoming constant. So the only solution left for the body is . . . total surrender—total. And in that state of total surrender, it realizes that that vibration is—how can I say it?—that vibration is not a vibration of dissolution, but something—what? The unknown, completely unknown—new and unknown. Sometimes it panics. And it can't say it suffers much, I can't call that suffering; it's something . . . quite amazing. So the only solution for it is to nestle in the Divine: what will happen will happen.*

I so profoundly understood what She meant, and I said, "Yes, the 'other thing' must be so different that it's like death for the body!" *At least, it's the equivalent. Yes, exactly. But . . . [and She smiled] it isn't fooled. It isn't fooled. It KNOWS that this is not what people call death. . . . But it's a funny life all the same!*

Then She laughed. *I'll soon be dangerously contagious, you know!*

This was in April 1970.

The mystery of the unknown.

*Something the cells do not yet understand, but they know, they feel. They feel as if thrust by force into a new world.*

It is matter finding matter's key.

The exit from the mineral, vegetable and animal kingdoms.
The beginning of the supramental being.
The exit from the second web.

NINE

# Innumerable Life

Indeed, it was a mystery, deeper than the appearance of mental man among the animals, deeper than the appearance of any species among other species made of the same substance; we would have to go back to the explosion of life in minerals. It was as if another life were being born. A little like: How did the mineral learn how to "live"? It was the shattering of its peaceful solidity, a wave of "something" indistinct, impalpable, "unreal" and supernatural, or of another nature, suddenly seizing its crystals and atoms, crushing and disintegrating them— an enormous disintegration. But what about reintegration? "Life" was perhaps something invisible to the mineral, too rapid for it, and it could only feel its disintegrating effects. But to what degree is that supramental life accessible to us thinking men, firmly settled in our cultured molecules? We can see the disintegration all right, but, even if our interest extends beyond the narrow, or broad, range of a mental life seeking economic, political or esthetic panaceas amidst the bankrupt little minerals, how can we expect to participate or collaborate in that supramental life, which, after all, seems to mean the dissolution of all our good life, perhaps even the dissolution of these fair bodies, if the experience of the human prototype called Mother is any indication? And who would ever have the courage to go through such an evolutionary ordeal deliberately? True, Mother's experience was rather "concentrated," as if centuries had been packed into a few years or a few months, and we assume that the experience will extend over several generations, whose cells will become increasingly refined, clarified, plastic; we also assume that the great supramental wave will seize (or is already seizing) all those bodies despite themselves, whether they deem it right, wrong or indifferent, and work in them underground, beneath the web of the physical mind, at cellular level, wearing down and undermining and pounding that fortress of death and disease and inflexible laws, to make them suddenly or progressively emerge into that new matter, freed from the mental tyranny—a life already singularly lighter, with a beginning of little song inside. But what would ultimately happen to this body of the old kingdom? Final disintegration or what? Does one pass into the invisible and unsubstantial (for us—even if it is another kind of substance and another kind of visibility) or what? But after all, the vegetable and the animal remained quite substantial and visible after the explosion of the mineral, and there is no reason to believe that evolution will be less reasonable than before, even if it does disturb the reason of the metal. Who knows, we will perhaps grow other eyes? Perhaps we have yet to know the earth's entire spectrum of visibility—all its naturalness, so to say—and that it is not forever confined within the range of a spectrograph? What is going to happen?

This is a little the question Mother's body poses for us, as if we were before an accelerated, advance specimen. Evidently, that body could not be "accelerated" without our also feeling the acceleration—a "dangerous contagion," as She said, laughing. It is truly as if, for once in its long history, evolution presented us its future data in a body, in the flesh. Will it survive or not? Will it succeed or not? Where are we going?

She wondered, too.

## The Son of the Cells

The heart of a tree is soft, then it lignifies: year after year the concentric rings of its "age" are formed. We could probably observe the same process, or a similar one, in the layers of mother of pearl or coral and at every level and in every branch of evolution, from the atom weaving or attracting its electrons around its nucleus, right up to the sun and its planets. But it always results in that hardening, that calcification or agglutination, and we cannot help thinking that what we call "matter" is not the original phenomenon but an induration. "Matter" is a solidified habit. And it seems that the supramental being follows the same process of concentric formation as all other bodies, but without induration. The primary substance is gradually agglutinated around the pure little vibration of the cells, is slowly deposited around that nucleus of call or prayer, or love perhaps, endlessly repeated by the mind of the cells. But, while in the beginning of life a crust or some kind of shell was formed in the course of time to differentiate beings, forms, various modes amidst that mass of amorphous Consciousness-Force or Consciousness-matter and to protect that precarious life, here, at the other end of evolution, the individual is already formed—that was the goal of this long journey—already *cellularly* conscious, and because he is conscious and individualized, he can deliberately amalgamate the primary substance without any need to lignify it or form a crust to distinguish and protect himself from the rest of the universe. The great Fear is gone. His protection lies in the very density of that substance of consciousness: every time, Mother's physical cells felt about to burst or vanish into that flood of solid power. But, as it is no longer confined in a fixed and rigid form, that body—the new body—can merge with everything and circulate everywhere, in everything. It is innumerable life. It is *physical* life without partitions. It is the great oneness of life—oneness of matter, oneness of consciousness—experienced materially. All those painful millennia of separation in a little prison of fossilized matter were preparing the wonder of that life. And one understands why it was necessary to become *cellularly* conscious in order to establish that life. It is evolved, conscious, individualized matter that builds its own pure body with the very substance it started from in the early, amorphous ages.

A son of the cells.

For let us make no mistake: that body is not built by the mind; it is created neither by the heart nor the feelings nor spiritual concentration—we all have a body of consciousness, and that's an old story, even if we are too ignorant or too blind to notice it with our eyes hampered by darkness. When we are in the least evolved, conscious, developed in our mental consciousness, when we no longer just shovel up "ideas" but are able to concentrate around ourselves and wield a mental, emotional or spiritual force, a subtle body is formed—a mental body, a body of consciousness, a body of energy, we could say, made up of the sum of all our vibrations—in which we travel. This is the oldest story in the world. It is in that subtle body that Mother used to meet Sri Aurobindo in Paris, in 1903, without even knowing him. It is in that body, as I related before, that I traveled 6,000 miles to witness the suicide of a friend in a room I did not know, in a town I did not know, but which I saw and was able to describe as accurately as if I had been present physically. This is also an example of the spinning of forces: We spin and churn out mental, vital or psychic forces the way others churn out layers of calcium. And a "body" is formed. But this is not at all the body that concerns us in Mother's case, not at all a "subtle" body; it is a material body, created by material, but conscious, cells. Only, what is involved here is not matter or the level of fossilized matter we know. It is primary matter. The cells alone can "understand" and recognize that matter. The first time Mother discovered the *material* world where the living and the "dead" were together without distinction, it is the body, the cells, that perceived the existence of that world—which Mother, who had every conceivable vision, had not seen in eighty years (1962). The mind only lives in its head, and our microscopes are just the enhanced eyes of that head. But the cells know. They are just reaching the point of their evolution when, free from the cage of the physical mind—that first body of fear—they will be capable of having direct knowledge, in their own way, of the material universe and of finding their own means of locomotion. What is being formed is a new species, obviously with a new type of perception, but it is nevertheless a very material species—probably even more material than the one derived from our mind. It is the supramental species.

And finally, it is only right and logical that an evolution of matter result in a blossoming, a flowering of matter itself, and not in the triumph of a little clown in his mental cage.

Then we will realize that we knew nothing of the world.

## Cellular Ubiquity

For a very long time, Mother did not fully comprehend what was happening, and we now understand more and more why Sri Aurobindo did not tell her anything, why He had not divulged his secret: the body itself had to find the way, *create* the way. Explanations are only meant for the mind, but the mind has nothing to do with this. On the contrary, it may take its fantasies for realities. For the body, there is no such thing as "fantasies"; it can only understand what it experiences. At times, however, strange experiences would literally burst in, the way we sometimes emerge unexpectedly into a clearing, then the green curtain of the forest closed again, and the long, slow, blind progress in which nothing seemed to happen for years would resume, only for the experience to come back again in a sharper, vaster, more precise way, as if Mother had been advancing in a subterranean way without even always knowing that this thing was tied with that one or was its continuation. The new body was formed quite invisibly and slowly around the cells, one thin layer or coat after another, with each of those calls, those silent prayers, those vibrations of aspiration amidst the great collapse of the old body.

One morning in 1962, just a few months after the experience of the great "pulsations," after her exit from the first web of the physical mind and when Mother had completely stopped leaving her room upstairs, I found her with a sort of puzzled, bewildered look on her face, like somebody confronting a perplexing riddle. *Some strange things are happening.... I don't know if you know the difference between the memory of an inner experience (in the subtle physical or the subconscious, in any inner realm) and the memory of a physical fact? There's a great difference of quality between the two. There's the same difference as between inner vision and material vision. Material vision is precise, with well-defined limits, and flat at the same time (I don't know how to describe it: very flat and superficial, but very exact—the kind of exactness or precision that defines things that are not at all defined). Well, the difference between the two types of memory is of the same nature as the difference between the two visions. And I noticed these past few days that I remembered leaving my room and going downstairs, seeing people and things, speaking, organizing certain details: several different things belonging to the PHYSICAL memory. Not at all things that I saw while exteriorized, with the inner vision—the MATERIAL memory of having done certain things. Afterwards, I had to recall them as one recalls a memory, and that's what suddenly drew my attention. I said to myself, "But did you actually go down there materially?..." Everybody is here to prove that I didn't go down, that I didn't move from here.*

*Yet, I have the* MATERIAL *memory of having done so, of doing other things, and even of going out! . . . Well, I have a real problem on my hands. Not only is the memory of it absolutely material, but the effects of what I said and did* EXIST. "Were you able to check that actual changes occured as a result?" I immediately asked Mother. *Yes, they occured! I had said, "This must be done that way," and it became that way. For instance, if I said to someone, "Put this there," that person put it there. She doesn't realize I was the one who told her, but she did it (she doesn't realize because she doesn't have the same consciousness as I). Furthermore, the effects of my intervention occurred even before I remembered it. It happened in reverse, you see: when I noticed that such and such thing had actually been done, I said to myself, "Well, this person is really marvelous!" and then I suddenly realized: "But I am the one who told her to do it!" I am the one who told her. Then only came the recollection—the "recollection," not the memory of a vision, but the memory of something that you've* DONE. "Isn't it an exteriorization in the subtle physical?" I asked. *No, not at all! Because the memory of an exteriorization in the subtle physical is very different. I have a lot of experience with that, you know, for about some sixty years! I know that phenomenon. If you will, this is exactly the kind of experience one has in the physical Falsehood, in the ordinary physical consciousness.* "Could it be material bilocation?" *It's possible . . . It may be that. . . . Ubiquity or something of the sort.* But even after putting a label on the phenomenon, Mother was not the better for it.

But the other persons who, following Mother's instructions, had made the necessary material changes had no physical recollection of having seen Mother. . . . Strange. Yet they had actually made the changes. How is it possible? This is Mother's answer: *For them, whenever they have experiences (they have no knowledge; ignorance is the most widespread of things), they regard everything as dreams. So it's a waste of time to try to ex-*

*plain to them—they won't understand. Everything is just dreams, dreams, dreams.* Then *two* physical realities? Or just one with two levels of matter, two ways of experiencing the same matter separated by the barrier of the physical mind? A true physical, as Sri Aurobindo said, and the other? True matter, matter as seen and experienced by the body's cells, and the other as seen and experienced by the mind? But for the ordinary consciousness living in the mental cage, everything happening on the other side of the cage, yet in the same matter, is like a "dream," another "world." If asked why they made those changes, they answer: I thought, I felt; or else: I had a dream in which Mother told me... In other words, the body of the cells is made of a more refined degree of matter than our own substance; there is no shell over it. Our present eyes only capture the opaque shell of things. And yet it is a material body materially moving about in our material world.

But Mother was still facing her problem and all the "ubiquities" in the world could not resolve anything, except that as usual a nice label had been added to camouflage our ignorance and tame the unknown. Our world is plastered with labels, in Greco-Latin to boot.

This was in 1962. One year later came a sudden, unexpected explosion in the forest, incomprehensible as usual, but more "situated" this time: *There must be something new in the consciousness of cellular aggregates—something. As a result, I had a series of fantastic cellular experiences, which I can't even explain and which must be the beginning of a new revelation. When the experience began, something in me was watching (there's always something in me that watches in a slightly ironic way, always amused)* ... [that's exactly Mother], *and it said: "If this were happening to somebody else, he'd surely think he was seriously ill or half crazy!" So I stayed very quiet and said: "All right, let it be. I'll just watch and see what happens."* ... *Indescribable! Indescribable! The experience will have to be re-*

*peated several times before I can understand. Fantastic! It began at half past eight and lasted till two-thirty in the morning, which means that I didn't lose consciousness for a second while I observed the most fantastic things. I don't know where this will lead. . . . It's indescribable. You see, you become a forest, a river, a mountain, a house—and it is the sensation OF THE BODY! It is the altogether concrete sensation of this [Mother pinched the skin of her hand]. And many other things. Indescribable. . . .* Again, I brought out my little label and asked Mother, "Ubiquity?" *Oneness,* She answered. *The sense of oneness.* Naturally, at the cellular level, everything is ONE; the great oneness pervades everything, without separation, without a shell. Even mountains are free of a shell. Only men have a shell. And She added this, which is rather mysterious, or seemed so at the time: *It is obvious that if this becomes a natural, spontaneous—and constant—thing, then death can no longer exist, even for the body. . . . There is something I FEEL in this without being able to express it or quite understand it mentally. There must be a difference, even in the behavior of the cells, when one leaves one's body. Something else must happen.*

What happened in 1973?

What is happening now?

A body of the cells that does not die. . . .

It was 1963.

Then the green curtain of the forest closed again for six years. And Mother kept repeating to me, although I did not really understand, *It's the CONSCIOUSNESS of the cells that must change.* I could not quite understand how the consciousness of the cells would modify the body's processes, and She did not really know either, but She knew it was the key to the mystery, and She worked and struggled to awaken that mind of the cells, the pure little vibration in the depths of the cells, to clarify and free all that substance from the old hypnotic remnants of the physical mind. And gradually, She felt something forming within. She felt a kind of explanation without words rising blindly from the depths of the body: *What takes time,* She told me in 1966, *is to prepare matter, this cellular matter as it is presently organized, to make it flexible enough and strong enough to withstand the divine Force.* . . . Indeed, there was that tidal wave of power which came in small doses as the web gave way, that "boiling porridge" of primary matter, those experiences of "dissolution," until She could withstand the "golden onslaught," as Sri Aurobindo said, when the second web finally came unraveled. *It takes a very long time. But it explains everything—absolutely everything. The day we can describe it in detail, it will be quite interesting.* She would never explain it; She would live it. It is I who am trying to untangle the vines in the forest, hack a way through the dark, green curtain, link unexpected clearings together. *There's a small inkling of what the being Sri Aurobindo called "supramental" will be, the next creation. A small inkling. And as Sri Aurobindo said, it's an explanation that comes from inside out—the outside, the surface, has only a very secondary importance and it will come at the very end, when everything else is ready. But it's from the inside out, and it begins in a rather precise and interesting way.* . . . *A lot of time.*

From the inside out—like the butterfly inside the caterpillar.

Then suddenly, in 1969, just six months after the great turning point of 1968, when Mother's body was left on its own with that mind of the cells, which had no choice but develop since it was the only mind left, came a second explosion, a radical one this time: *It was—never, ever has the body felt so happy! It was the complete Presence, absolute freedom and a sense of certitude. [Its death] didn't matter in the least: these cells, other cells—there was life everywhere, consciousness everywhere. It was absolutely wonderful. It came in without effort; it went away simply because . . . I had too many other things to do. And this is the DIVINE SENSE, you know, it means having a divine sense. During those few hours (three or four), I absolutely understood what it meant to have a sense of the divine consciousness in the body. And whether it was this body, that body or another body [Mother made a gesture all around her, indicating the body of this person, that person] didn't matter at all: it just went from one body to another, totally free and independent, with the full knowledge of the limitations or possibilities of each body. Absolutely marvelous! I had never, ever had that experience before. Absolutely marvelous. It went away because I was so busy and . . . And it lasted several hours. Never has this body, in the ninety-one years it has been on earth, felt such a happiness: freedom, absolute power, and no limits—no limits, no impossibilities, nothing. It was . . . all bodies were this body; there was no difference.*

Innumerable life. Innumerable *material* life.

No more shell, no more prison.

And a life on earth.

In an earthly body formed by the cells.

A body that does not die.

## The Fact of the Corpse

All of a sudden, Mother regarded death differently—but not for long. One month after that experience, Mother thought She had found the solution (indeed, it may have been *part* of the solution), and She said this: *One question kept coming back: All this work of transformation of the cells, of the consciousness in the cells, seems to have to be wasted since the body will disintegrate. Then, clearly, almost concretely, there came: There is a way, which is to prepare within oneself, before death, a body with all the transformed, illumined, conscious cells—to assemble and form them into a body with the maximum number of conscious cells—and once that is completed, the consciousness enters that body and the other one can dissolve—it doesn't matter anymore.*

This is simple indeed.

But it does not seem to make evolutionary sense.

One simply discards the old rag.

A rag that has painfully prepared that pupation year after year, millennia after millennia.

There *must* be something else.

There must be a missing, living link between that other body and this one.

Is it a transformation of matter or a dissolution of the old matter?

But then, where is our own meaning in all this?

*There is a sense of fluidity, of plasticity that is becoming more and more evident for the consciousness, with just—just—something on the outside that . . . that feels more and more like an illusion. It feels somewhat like a piece of bark clumsily covering certain places.*

What is going to happen to the "piece of bark"?

That is the mystery of the last three years of Mother's life. An increasingly acute, painful, almost excruciating mystery, until that day in 1971 when She exclaimed, *The inner consciousness can say, can be conscious that that suffering is unreal, but the physical consciousness can't! It can't—something HAS to change. It isn't a matter of entering a consciousness in which this physical consciousness ceases to exist—it must change, it MUST change. . . . The FACT must change, you see. In order for the transformation to be genuine, the body TOO must attain a harmony above—above all diseases, above all accidents.*

The Fact must change.

What is going to happen?

*The surface, the part that feels like a piece of bark, will be the last to change. What is going to happen? I don't know . . . I don't know. But it will be the last to change.*

It will change, She said.

That is the whole mystery.

The mystery is no longer that of forming a new body in evolution; it is that of transforming the old body—the link between the two.

Or no link?

It seems as if this old matter remained our touchstone.

The fact of the corpse.

The curtain of death has to be pulled open in all bodies.

TEN

# Victory over Death or in Death?

It may be our human sentimentality that makes us wish for the transformation or glorious blossoming of this old body. Evolution is not sentimental and it has often demonstrated it could drop intermediary species or let them die out, or else remain a certain time until all evolvable elements have reached the next higher level. And if the evolutionary goal is indeed to evolve that new "innumerable" body, why should it remain attached to this old form once the latter has discharged its duty? Sri Aurobindo did foresee that humanity would not rise "in mass," all at once, to the supramental level, and that the two levels might exist side by side for centuries (probably within two worlds separated by a veil of unconsciousness or "death"), until all the capable elements have crossed the threshold. Thus, gradually, century after century, all the species would slowly climb the ladder of consciousness or stay in their stagnant but harmonious perfection. What is not known is to what degree, perhaps quite devastating and unexpected, the formation of that new body or several new bodies endowed with that rather fantastic vibratory power—since it is the very power that drives the atoms, the innumerable "gold powdering" that moves freely in everything, through everything—will change or upset or accelerate the present evolutionary data, in the process toppling a number of perfectly scientific walls of impossibility. It is even what is happening now. Today's insurmountable is perhaps a mere breath that will leave nothing but a smile behind. But we are concerned with the transition, our transition, and we wonder about two things: first, why would these physical, animal cells—which have toiled so long, which have become conscious and sent out their little distress signals or calls, developing a mind of the cells as a tireless vibration of prayer or joy or pure love, a tiny nucleus of gold, a little powdering of the innumerable golden powdering of the universal substance, spinning and weaving that substance around themselves—why would they not permeate all the old body, not change this old substance, forcing it to operate differently and clearing up this opacity? Why should they go on decaying? In fact, if they are truly conscious, they *cannot* decay. Only what is unconscious dies. Death is the ultimate sign of unconsciousness. Therefore, there is not any "theoretical" impossibility to the old animal body being transformed, to an evolutionary continuity which would justify our long struggle from one human or nonhuman cradle to another. It would be a "visible" proof.

There remains to see what the practical difficulties are.

But there is also our second question, which invisibly hinges upon the first one. Everything hinges upon that famous "visibility." We cannot help looking, or trying to look, at the next species with the eyes and understanding and sentimentality of the old species. If we were equipped with animal organs enabling us to see that primary substance, actually to witness the formation of the new body within the old one, to watch it slowly take shape, move independently and radiate its joy and beauty—to what extent would we be dependent on the old masquerade, even if it made the birth of that beauty possible? We would just have to let go of the old hide. And the other one would be right there, visibly, gloriously.... In fact, we do have those organs: the cells see and know. We have thousands and billions of cells, just covered over, veiled by the web of the physical mind, which superimposes its sad and painful reality, its old habit of suffering, falling ill and dying, its innumerable gray illusion that wraps each gesture and each footstep in fears, apprehensions, laws, legal catastrophes, like an octopus. It is the veil between the two worlds. It is truly the veil of death. So we say that one side is the "concrete," real world and the other is the impalpable, "invisible" world, a physical world perhaps, but "subtle" nonetheless. But what would happen if that veil, that opaque periphery wore away and disappeared, if our cells actually saw the reality? First, it would likely result in an overpowering desire for transformation in all human bodies (not to speak of external upheavals), as if for once they breathed real air—they would not want to breathe any other air. But many might find that air quite unbearable for the thick layers of filth covering them. This is perhaps why the wearing away of the veil is slow and merciful and chaotic. For indeed there are no elect; the whole evolution is elected. And we always come back to these two worlds side by side, one in the other, these two types of humanity, we could say. One would slowly open the eyes of its cells, slowly weave a new body of joy and beauty, while the other, dilatory, would also be led and driven, by its sheer suffering and chaos, to seek something else, to want something else, to open matter's little eyes in its own cells. And when everything has reached the point of homogeneity, there will no longer be "this side" and "that side"; everything will be on one and the same side. The line of demarcation is the body's death. It is the point where old matter no longer follows the movement of new matter and fails to transform itself. So it sheds its skin and merges with the other. But this is the very essence of falsehood. The corpse is the testamentary residue of Unconsciousness. It is the very sign of the veil. And one wonders if it will ever be possible to enjoy a complete and real life on earth, genuine, totally unveiled, so long as death is not overcome *there*, changed *there*, in its very nest of Falsehood—because *there* would remain the very origin of the veil. What appears to us as the last deception, the ultimate masquerade may just hold the key to the ultimate unveiling and absolute fulfillment. We actually must change the death of matter. Skipping over it into a new species is not enough. The veil separating the two worlds must be undone in the body.

One morning, I saw Mother arrive with four lines from *Savitri*, "The Debate of Love and Death":

> *The great stars burn with my unceasing fire*
> *And life and death are both its fuel made.*
> *Life only was my blind attempt to love:*
> *Earth saw my struggle, heaven my victory.*[1]

And Mother had a special kind of light in her eyes, as if She had received the message, met the message. *Savitri* says, "*Life and death are the fuel.*" Then she says, "*In my blindness, LIFE ONLY was my attempt to love.*" She doesn't say, "Life was only . . ."; She says, "Life only was . . ." "Because my attempt to love was blind, I confined it to life—but I have won the victory in death [that is, in 'heaven']." It's very interesting.

*Earth saw my struggle, heaven my victory*

"But it's the earth that should see the victory," I said to Mother. "The victory should take place on earth, shouldn't it?" *Yes, but Savitri couldn't win the victory on earth because she lacked "heaven"—she couldn't win the victory in life because she lacked death, and she had to conquer death in order to conquer life. That is the idea. Unless you conquer Death, the victory cannot be won. Death must be overcome; there should be no more death. It's very clear.*

Then She added, *According to what Sri Aurobindo says here, it is the principle of Love that changes into flame, and then into light. It isn't the principle of Light that changes into flame by materializing itself; it's the flame that changes into light. The great stars give off light because they burn; they burn because they are the result of love. . . . It is my experience of the "pulsations." The last thing—beyond light, beyond consciousness, beyond . . . —the last thing one comes into contact with is Love.* [i.e., the supramental fire, the "golden onslaught."] *According to the experience, it is the last thing to manifest, now, in its full purity, and it is what has the power to transform. That's what Sri Aurobindo appears to say here: The victory of Love would seem to be the final victory. He said that Savitri was "a legend and a symbol." He is the one who made it into a symbol. It's the story of the meeting between Savitri, the principle of Love, and Death. And it's over Death that she won the victory, not in life. She couldn't win the victory in life if she hadn't won the victory over death. It's very interesting.*

One cannot skip over death.

Years earlier, in 1963, while speaking of death, Mother had told me, *It's almost as if it were THE question given me to resolve.*

How was She going the resolve the problem?

The transformation of this animal, physical body is the very symbol of the victory over death. It is the very knot of death.

Is it necessary to die in order to overcome death?

How can one win the victory in death and win it in a body at the same time?

Or else one must die, undergo death *while in the body,* and come back victorious.

It is a mystery.

Mother did say, "Victory IN death."

This could well be the "dangerous" mystery with which She had been struggling since 1970.

Perhaps She is still struggling with it.
But death has to be overcome in a body.
Once the veil is pulled away *there*, it will be pulled away for all bodies.

*Death must be overcome; there should be no more death. It's very clear.*

ELEVEN

# Transformation

For a long time, and perhaps till the end, Mother did not know the path. If I may say so, this is perhaps the first time that, with this pen, the path is being outlined—bodies do not need to know; they simply walk. It is our heads that need to know, that is, to have a map to follow. But a map is of no use to the body. The new world is just this very step; the map is just that aspiration. And all bodies that want to follow the path will need only that aspiration. Actually, that has been the only way since the protoplasm—an aspiration. And it will always be the way. We meet at the end that with which we began. We constantly carry the goal with us; the goal is there every minute. When we realize that, time vanishes: today is millions of years old. If our head realizes it, we drift into contemplation, soar into a white eternity, and nothing is changed. When our body realizes it, we perhaps touch one of the keys to the transformation, because time is the enemy of the body; we enter a golden eternity which seems to have very particular properties. The body is what holds the key to the long journey. So it does not need any map; it needs to be. And what is most remarkable is that when the body finds that eternity, it is no longer felt as a static eternity like that of monks, poets or meditators, but as an eternity that seems to be filled with supreme dynamism—an eternity in lightning-fast motion. It is another kind of time. And it is not eternity as we know it. There exists a "time" of the body which perhaps holds the secret to another space, which perhaps holds the secret of matter, because matter is already a sort of coagulated time. And which perhaps holds the secret of the dissolution of death, which goes with time. In the face of Death, we have found only a white eternity: two kinds of dissolution, one in white and another in black. When we find the third kind of time, the time of the body, the time that neither dies nor sinks into a rut of eternity, we will perhaps have the complete key. Only the body can know.

Mother's body was slowly emerging into a third kind of time.

## Transformation or Change of Perspective?

It was a sort of living contradiction, as if Mother found herself not before but *in* two totally different paths at the same time. Our head has no trouble following two contradictory ideas and living more or less comfortably in its own confusion, but how can a body have one foot here and one foot there? Two paths or two bodies that seemed to go in opposite directions, or maybe parallel directions meeting at infinity—but infinity is pretty far away. On one hand, She vaguely, imperceptibly felt that new body being formed in her. *I am not entirely sure I don't already exist PHYSICALLY in a true body,* She said as early as 1963, just there, in matter, on the other side of our cage. *I say I am not entirely sure because the external senses have no proof of it, but . . . But from time to time there comes a sort of compelling sensation: for a minute, I see myself, feel myself, objectify myself as I am. But it only lasts a few seconds and, prrt, it's gone—to be replaced by the old habit.* Suddenly, we wonder if this overwhelming, unimpeachable "scientific" materiality of the world is nothing but a habit, an old habit of seeing things in a certain way. *People who see me at night, those who have that sort of vision in the subtle world do not see me like this* [Mother pointed to her external form]; *they see me as I am, and they say so. They say: Oh, you're like this and like that!* . . . And She added, *But for one to take the place of the other?* . . . That was the whole question. It was the whole ambiguity: was the one going to take the place the other, or was the other going to change into the one? She did not know. She went from one to the other. It was a sort of hellish coming and going between the two—and evidently the key to the solution was in that very hell, the possibility of joining the two was in that very contradiction. It is easy to say but not so easy to live. In fact, it was unlivable, except for her. Then, one day in 1970, things crystallized, and She herself saw that body: *Well, I saw my body! What it will be like. It's quite good* [She laughed]. *It's quite good. A form that looked like our own body, but without sex, that is, neither man nor woman. It is not such a different body, but so refined, so . . . something so refined. It had a color . . . somewhat like Auroville's color [orange], like that, but vibrant, I mean as if it were . . . not luminous, but with a sort of luminosity. And I was wide awake. I wasn't asleep; it was not a dream. It was as objective as waking reality.* And I was reminded of her vision, twelve years earlier, of that supramental ship carrying tall beings of orange color. A long journey. As if what She had seen then, far away in a vision, had come into matter—except that it had not really "come"; there were just layers of opaque consciousness to cross, and it was merely "at the other end," "over there," on the other side of the sleep of our physical mind. Once the veil is crossed, it's *here,* with our eyes wide open, as "objective" as the table and the chair, and even more objective because its degree of consciousness is greater than a table's or a chair's—it is denser. I think I even understand what Mother meant when She said that that new matter is "without division." Our scientific matter is full of empty space; it is made of an immense fragmentation of particles at enormous distances from one another (enormous for that scale*). While that matter is compact, dense, without "gaps." "As dense as the diamond, but elastic," She said in 1906.

It is strange to recall that the first revolution or revelation in Mother's life occurred around the age of eight or nine, when She was told, *"You see this table? You think it's a table, that it's made of solid wood—it's just atoms moving around." ... It caused a sort of revolution in my head, and a sense of the complete unreality of all appearances. Suddenly, I said, "But, if that is so, then nothing is true!"*

Now She was in the midst of her second revolution, behind the other "surface," the atomic one.

We have never found what is behind the atom. Where is the ultimate, indivisible nucleus?

And Mother wondered what was going to happen between the old body and the other one. Will the old habit of looking at things not change in humans? *Will there always be a world as it is now? ... One can well conceive of a world where people would live in that new state, which would develop according to its own particular laws. But would the existence of such a world negate this one? ... This, you see, is a question that has yet to be answered.* As early as 1962, She was wondering whether the whole problem did not, in fact, come down to a sort of collective change of perspective, something that would cause the experience to spread, for we always forget that nothing is separate, no experience is isolated as though walled off, and the earth is one single body of evolutionary experience. *Everything*

---

* "Like a Ping-Pong ball in relation to Houston's baseball field," said the American scientist Robert Jastrow when speaking of the distance between the nucleus and the electrons of an atom in *Red Giants and White Dwarfs: The Evolution of Stars* (Harper, 1967).

*almost comes down to an ability to spread the experience, or to EMBRACE the world within the experience (it's the same thing). You see, we must forget there's this person, that person, this thing, that thing. . . . Imagine—if you can't experience it concretely—imagine that there's only ONE exceedingly complex Thing, and an experience occurs in one point, spreading or taking in the whole, depending on the case. That's the only explanation for the "contagion"—Oneness.* A sufficiently contagious experience. But this is the whole experience of Sri Aurobindo and Mother: the collective wearing away of the veil of the physical mind. If it wears away in their bodies, it must wear away in the body of the world. And at times, Mother thought She had caught "the tail of the solution," as She called it: *A sort of knowledge (is it a knowledge?) or foreknowledge is given to the body of how this appearance will change. It seems to be very simple and very easy, and it can be very swift, for it isn't at all—it will not AT ALL happen in the way people think or expect. It will rather be like the vision of the TRUE internal movement IMPOSING ITSELF in such a way as to veil the false vision. . . . There is something true, the true physical, which is not perceptible by our eyes as they see now, but which could make itself perceptible by INTENSIFICATION.* That is what Mother always said: It is not as if we had to create that other world from scratch; it is already there, totally existent, "and it would just take a little spark," an invasion of the Real. *And that intensification is what would produce the external transformation, replace the false appearance with the real form. . . . But when you say that now, people visualize a "psychic" vision or a mental vision—that's not it at all! I don't mean that. I mean a PHYSICAL vision, with these eyes* [Mother touched her eyes]. *But a true physical vision instead of the distorted vision that exists now.* A physical change of sight? We cannot help thinking (we = the fish in the bowl) that that "other" world is "another" reality—but it is the *same,* with different eyes! It is not supernatural; it is a more accurate natural. *Basically, Mother concluded, it means that true reality is far more marvelous than we can imagine, because what we imagine is always an improvement or glorification of what we see—but that's not it! That's not it. We can only think of things changing from one to another, you know: regaining youth, eliminating all signs of old age, etc.* [in short, the improvement of the fish inside the same bowl]—*it's the same old story. That's not it.*

That is very simple indeed: a worldwide stroke of a magic wand. And it is quite possible that things actually will happen that way, to a degree. It is quite possible that all the Falsehood of the world will be suddenly filled with black or gray, vanish from our sight. . . . But if the only measure and only weight are the weight of true consciousness, we wonder how many humans would suddenly "vanish" in the process. *I don't at all know whether the false appearance would not continue to exist for those who would not be ready to see the real thing,* Mother wondered. *In any case, that would be an intermediary stage: those who have their eyes open (what is called "eyes open" in the Scriptures) could see, and they could see not by effort or endeavoring to, but because it's self-evident, while those who do not have their eyes open . . . At least for some time, that's how it would be: they wouldn't see. They would still see the old appearance. The two could be there simultaneously.* It is clear that if only a few hundred or few thousand humans were "taken into" the new vision by the contagion of the new world, it would have incalculable consequences on the rest, which could not help but be "affected" by that other human "species" whose superior traits and superior joy and superior level of harmony would pose quite a formidable question, perhaps even a challenge, an invisible constraint for the less evolved, increasingly distraught in their suffocation. Clearly, this would be a similar, and perhaps more powerful, phenomenon as the influx of the first wave of amphibians among the fish. It may be the phenomenon happening now, the invisible, less and less invisible contagion.

But...

There is a rather formidable "but."

The appearance of the world changes for those who have their eyes open. Even the appearance of the body changes: the true light, true consciousness, true essence of things asserts itself. And there is no doubt that the more obvious and visible the true vibration becomes, the more it rectifies, almost automatically, the false, deceptive vibration—the more we will be *forced* to be true. There will no longer be any possibility of cheating. A momentous step will have been taken *in the consciousness,* with its incalculable consequences for the harmonization and organization of life. But there still will be a body to be buried.

That is the crux of the problem.

An aging, shriveling, dying body.

The very symbol of Falsehood is still there.

Mother looked at that increasingly perceptible new body and at that increasingly pressing death in her old body. What was going to happen?

When all is said and done, the FACT must change, She said.

Can it change?

Mother was as if on two paths simultaneously: a path of life and a path of death; an imperishable body and a decaying body. She was at once in life and in death, on both sides of the bowl ... as if both had to become one or change into a third thing.

The body is the bridge.

## The Problem of Transformation

It is hard to imagine how that "vibrant body"—dense but fluid, slightly luminous, without structure as we know it—could pass into or enter this old body equipped with a skeleton, whose very opacity seems to be the sign of its existence. Undeniably, at the beginning of the primary era, it was impossible to imagine how those metals and rocks could ever change into forms that were living and "fluid" compared to that completely inert original opacity. The principle of it is therefore neither improbable nor even antiscientific on the evolutionary scale, however it may appear to us today. But in practical, day-to-day terms? Evolution does not begin tomorrow; it is done day after day, and even when a sudden mutation occurs, that mutation is the result of a long preparation, an accumulation of factors that suddenly provoke a breach and a change of equilibrium. What is the process of that new transformation? There must be a rational process to it, even if it is slow, invisible, and follows a "reason" that obviously eludes us because it is the reason of the next being. If we had had to rely on the reason of the metals, we would quite likely continue to have a very metallic universe today. Mother is the logic of the future, which did not prevent her from handling today's logic with mathematical rigor. She always went straight to the point, coolly, to take the mechanism apart. The "mechanism" was her passion. *There is very little difference between a calf developing in a cow's womb and a baby developing in its mother's womb. There is a difference—the intervention of the mind. But to envision a PHYSICAL being, I mean, visible as the physical is now visible and with a similar density—for example, a body that wouldn't need blood circulation or bones—is a far greater transformation than the one from animal to man; it would mean evolving into a being that wouldn't be built in the same way, that would not function in the same way, that would be a densification or materializa-*

*tion of . . . "something." This has nothing in common with anything we've seen physically so far.* "One can conceive of a light or a new force giving the cells a sort of spontaneous life, a spontaneous energy," I said. *Yes. That's what I am saying. Food could conceivably be eliminated.* "But the whole body could be driven by that force, couldn't it? The body could remain supple, for example, and, while having a bone structure, be supple as a child's." *But a child can't stand up because of that!* She exclaimed. *What would replace the bone structure, for instance?* "The elements could be the same, but they would be supple: their firmness would not result from being hard but from the power of the light, no?" *Yes, maybe. . . . Supple, plastic—that's also conceivable. In other words, the form wouldn't be fixed as it is now. All that is conceivable, but . . .* "But one could view it as a sort of blossoming, a luminous self-expansion, like a lotus bud opening up. A bud is something hard. The light must have that power. And that would abolish nothing of the present structure." And Mother was like Saint Thomas: *But visible? Something one can touch?* (She, too, very much insisted on "concreteness.") "Yes, simply like a blossoming: what was closed opens up like a flower. But there's still the structure of the flower." And She shook her head: *I still lack experience. I don't know.*

To experience concretely was the only thing that interested Mother, and the moment I tried to speculate or guess, She coolly dismissed it: "That's poetry, vagueness," She said. In the nineteen years I was with her, I never once saw Mother being "vague." *I am absolutely convinced, because I have had experiences to prove it, that the life of this body—what makes it move and change—can be replaced with a force, that is, a sort of immortality can be created, and the wear and tear also can disappear. And that can come about psychologically, by completely submitting to the divine Impulse* [i.e., a willing automatism], *which causes you to have the necessary energy and do the necessary thing at each instant. All that is unquestionable. It's not just hope or fantasy; it's fact. The body just needs to be trained and slowly transformed, to change its habits. That's possible. All that is possible. But simply how much TIME will it take to eliminate the need (let us take just this problem) of a skeleton? . . . Yes, time was the problem. Because a supramental body suspended in a world that is not the earth can't be it, can it? Ah, indeed! . . . What I mean to say,* She added, *is that perhaps a large number of new creations will be necessary for this to happen. For instance, the transition from man to the supramental being will perhaps occur through many different intermediary stages. That jump is what seems enormous to me, you see. . . . I can very well envision a being capable, through spiritual power, through the power of his inner being, of absorbing whatever energies are necessary, of renewing himself and remaining young forever—that's quite conceivable—even having enough plasticity to be able to change his form if necessary. But the total, immediate disappearance of this present type of construction—to change immediately from one to the other seems to . . . necessitate stages. There may be intermediary beings which will not last very long, you know, as there were intermediary beings between the chimpanzee and man. . . . But I don't know. Something has to happen that hasn't happened until now.*

Something that eludes us and which is perhaps right under our eyes, incomprehensible.

In all our facts, we always forget the "something" that is not a fact of the past.

But that is not all. Mother saw the problem in its entirety: *My personal experience is like this: Everything I do with the Presence of the Lord I do without effort, without difficulty, without fatigue, without wear and tear, naturally, in the great Rhythm—but because it's still open to outside influences, the body is forced to do certain things that are not directly the ex-*

*pression of the supreme Impulse, hence the fatigue, the friction. ... You see, there has to be something capable of resisting the force of contagion. Man can't resist the animal contagion. He can't; he has constant relations with it. So how will that being manage? ... It would seem that, for quite some time—quite some time—he will be subject to the laws of contagion.* "I don't know," I replied in my innocence, "but that doesn't seem impossible to me. I feel that as long as that Power of Light is there, what could possibly contaminate it?" *But everybody would vanish!* Mother exclaimed. *That's the point, you see. When THAT comes, when the Lord is there, there is not one person in a thousand who doesn't find it terrifying. Not in the intellect, mind you, not in thought, but right in the flesh. So suppose—suppose that it happens, that one being is a condensation or expression, a formula of the supreme Power, of the supreme Light—what would happen! ... I've seen grown-up people come here (I've made the experiment: I charge the atmosphere, the Lord is there), well, I've seen forty-year-olds come in, and ... prrt, literally run away in spite of all social rules of politeness, and after having ASKED to come, mind you! Yet all the conditions called for decent behavior on their part—impossible. They couldn't. It was too strong.* "Well, that's the whole problem," I said to Mother. "Because I don't see the transformation itself as a difficulty. I rather think it's the difficulty of the world."

The difficulty of the old, recalcitrant species.

*If everything could be transformed at the same time,* She said, *it would be all right, but it visibly doesn't work that way. If one being alone were transformed ... that would be unbearable, maybe.*

Such is the problem of the transformation, which perhaps is not only and primarily a physiological and anatomical problem, but a global problem, because evolution is the whole world, from protoplasm right up to us. There is *no* material impossibility, *no* physiological impossibility, any more than there was in the age of iron or nickel—there is always, eternally, the difficulty of a past that does not want to die and hangs on to the old forms. To its cherished suffocation. The experience of Sri Aurobindo and Mother may be the most severe trauma undergone by the earth since the appearance of life. It is only a beginning.

Yet everything can be miraculous. . . .
If something yields in human consciousnesses.
A little fact of joy.
And we wonder if our bodies will not surrender to joy before we do . . . and take us by surprise?

## Cellular Stabilization

That transformation of the body is as slow and invisible as the formation of the new body, and it is, in fact, concomitant, the two phenomena being inseparable: what spins out the new substance, the primary matter, also produces a deep change in the cellular operations and in the material body-substance. But there the process is slower for obvious reasons: everything would break if the change were too abrupt. *To fall to pieces or to transform oneself is . . . almost the same process!* Mother exclaimed. One can thus begin to appreciate the hell of the last years, the dreadful paradox She lived in. "To come apart forward . . ." The process of transformation, therefore, is very simply the same as the process of formation of the new body. What seems to us extraordinarily complicated, a kind of incredible impossibility, has the simplest key of all: the mind of the cells is what draws or fixes the formidable Current of the primary, supramental substance; it is the *fixative* or connector of the supramental vibration. This is why Mother used to faint each time She attempted to come into contact with the Supramental by neutralizing the mind of the body. For it is the link, the bridge between matter as it is, set and hardened by evolutionary habits, and primary, supramental matter, fluid and vibrant and so overwhelmingly powerful, without "gaps." One evidently does not "fix" such a current with impunity. It means a kind of tearing down of everything by inches, so to speak. The first bodies to form—the atoms and particles—lost no time in building a crust with that substance, and in dividing it up by dividing themselves from the gigantic mass of magma; then everything became further encrusted over that first crust. It really takes a complete widening of the cellular consciousness—exactly the opposite movement of the original encrustation—to be able to withstand the formidable supramental tidal wave, with just the very slight vibration of the mind of the cells, the minuscule nucleus of prayer or call, fixing or spinning out whatever amount it can withstand. Hence, that mind of the cells is the key, the simple key to the transformation as well as to the new body—the whole problem was to reach it. That was Sri Aurobindo's "mathematical formula" when He reached the cellular base. In it He found the missing link between matter and the Supramental, and there just remained to "work things out" . . . cell by cell. *Mind, for instance, is everywhere,* He noted in one of the only revealing texts He left on the subject, which has been partially quoted earlier. . . . *And there is too an obscure mind of the body, of the very cells, molecules, corpuscles. . . . This body-mind is a very tangible truth; owing to its obscurity and mechanical clinging to past movements and facile oblivion and rejection of the new, we find in it one of the chief obstacles to permeation by the supermind Force and the transformation of the functioning of the body. On the other hand, once effectively converted, it will be one of the most precious instruments for the stabilisation of the supramental Light and Force in material Nature.*[1] This is Mother's very experience: Once that cellular body-mind was freed from the hypnotism of the physical mind, it began to fix the supramental vibration and form a new body. It began repeating the Mantra as imperturbably as it used to repeat its old catastrophes. It is the stabilizer. *That's why only that mind was left,* She said after the radical cleaning out of 1968. *Apparently, if you will, I had become an imbecile. I knew nothing. Then, little by little, little by little that mind started to develop. . . . Sri Aurobindo said that if the physical mind is transformed, the transformation of the body would NATURALLY follow. We'll see!*

And we wonder whether the formation of that new body is not just the first stage of the transformation, instead of the final stage as many think. Whether the two phenomena are not complementary. Because if the new body is left by itself, it is still something "suspended in a world that is not the earth"—one vanishes into thin air. *That's not it,* said Mother. But then, that new body—faster to develop precisely because it does not have to break up all the old crusts—is the one that must, or should, infuse itself into the old formation, as much as the latter can stand it, until the two have merged into one another. It would then result in something that is neither the old frame as we know it nor the weightless new body, something that would provide weight to the new body and a *new* materiality to the old one—and that would truly be the supramental body on earth. The old body would be the bridge.

The whole question lies in that "weight" or new materiality. There, we are completely in the unknown—perhaps the "dangerous unknown" Mother spoke of—and we will not really know until the thing is actually done. How could the mineral comprehend life's "materiality"?

## Cellular Survival

We might think that that weight is simply a question of vision, and that if human vision changes, what is weightless will start to carry weight. But there still remains an old rag ending up in a hole. And we do not know why, but we suspect that is precisely where the whole secret lies. Otherwise, why speak of transformation? The first time Mother went to "Sri Aurobindo's residence," in 1958, I asked her if that "other" world, that true "lining" of the earth was the supramental world. She answered this: *My impression is that the kind of life Sri Aurobindo has presently is not the full satisfaction of supramental life. That "other" world had infinity, majesty, perfect peace, eternity—it had everything. ... Maybe joy is what was missing. ... Of course, Sri Aurobindo himself had joy, but I had the feeling it wasn't complete. And that's why I had to continue the work. I felt it could only be complete when things were changed here.*

Changed here, that is, when the two sides are ONE.

When the other body and the earthly body are ONE.

What is the mystery that divides them?

Perhaps it was the mystery being "worked out" in Mother's body?

*We can't both remain upon earth; one must go*, Sri Aurobindo had said to Mother. And Mother had replied, *"I am ready, I'll go."* Then He answered me, *"No, you can't go. Your body is better than mine. You can withstand the transformation better than I can."* And still nineteen years later (in 1969), with I don't know what mixture of pain and anguish, Mother exclaimed, *Why? ... How many times have I asked myself since then. ... Why?* As if the transformation of that body were the condition for the two sides to meet, the place where the veil is torn. But what veil? What "mechanism"? And I asked Mother, "Does this mean that your presence here (on this side of the veil) could help Sri Aurobindo materialize one day?" *Yes, yes. He said that clearly (I had asked him): "I will only come back in a supramental body...." But the real question is about that supramental body. I don't know. What change in this body of matter would make it possible for the very nature of matter to* change enough so that . . . the other body can infuse itself here, into our earthly conditions. This change in the nature of matter is perhaps the crux of the story: a "hybrid" matter, as it were, halfway between the supramental fluidity and the opaque hardness of our earthly bodies? A new transparency of matter? A sort of new milieu to be created that would permit the two sides to join. "There is perhaps a passage to open (for Sri Aurobindo and that world to materialize)?" I asked Mother. *Maybe. . . . But this body has never had any desire or ambition to perform miracles—it isn't interested. It has seen a lot of miraculous things, but it has always felt that . . . that it was the supreme Lord doing them (it seems quite natural to it, in fact). But mental fantasies . . . Whenever they come, the body rejects them. It says: Thank you, I'm not interested.* Indeed, the point was not to perform a "miracle"; the point was to change matter itself, the body itself. The sole fantasies of the body are reality. It was perhaps in the process of making the "miracle" without knowing it. *It isn't in the least interested in all the things that people find marvelous,* Mother continued. *It wouldn't be surprised to see Sri Aurobindo enter this room one day—not at all. But it doesn't feel like . . . like doing it, you understand. It doesn't feel the need to impress people—not in the least. We'll see!* And She laughed. But I insisted, because I deeply felt it was not a question of miracle, or even of Sri Aurobindo's appearance, but truly of the earth's transformation—the transformation of the physical, material milieu which would permit the two sides to join, somewhat like when the transformation of mineral opacity gave matter its first eyes. And that same afternoon, I sent Mother a note with two lines: "Savitri seeks Satyavan in death—therefore Mother will bring back Sri Aurobindo?" *Something like that,* She answered.

"To bring back Sri Aurobindo" means to join the two sides.

And the question really narrows down. Everything revolves around a something to be transformed in the nature of this earthly body, this earthly matter, for the veil to be torn—and that veil is what causes death and what separates the two sides. It is in the body that death can be overcome, in the body that the veil is torn. It is the body that holds the key to the joining of the two worlds into ONE—it is the body that makes "Satyavan" come back. The mystery of the transformation is neither the mystery of a marvelous rejuvenation of the body, nor even that of a glorious change of body; it is the mystery of death in its den.

Mother's body is now in the very den of death. Has it failed? Or is it still engaged in some incredible task? "Earth saw my struggle, heaven my victory."

There is but *one* transformation to accomplish, and that is the transformation of death.

That is what Mother was slowly, dangerously approaching.

She was entering death alive.

And those cells—those tiny cells with their unflagging vibration of appeal, repeated thousands and millions of times, as many times as there are seconds in a day and a night—gradually acquired a sort of strange, indestructible life beneath the external shell, which seemed to disintegrate as rapidly as the cells were integrated. What were they going to do? Were they the site where one remains alive in death? The site of the ultimate mystery and perhaps of the transformation of death? The site where the supramental light is "stabilized," the light that does not die? In 1965, the first time Mother succeeded in piercing the crust of the physical mind and reaching the mind of the cells, She said this: *There's a whole work taking place to PREPARE the transformation. What could I call it? . . . A transfer of power. The cells and the whole material consciousness used to be controlled by the inner individual consciousness (psychic, for the most part, or mental, although the mind had been silent for quite a long time). But now, that material mind is being organized like the other—or all the others, I should say, the mind of each of the states of being. It's getting its own education, if you can imagine it. It is learning things and organizing the ordinary science of the material world. It's quite interesting.... You see, all the memory resulting from mental knowledge has long been gone, and I only received indications from above; but now, there's a sort of memory being built from the bottom up.... It's as if the governing center had been shifted: it's no longer the same thing that makes you act. By act, I mean everything, you know: moving, walking, anything. The center is no longer the same.* An automatic cellular life governed by the mind of the cells, by that little universal, indestructible vibration. And as I did not really understand, then, what that new mind was, I asked Mother, "How exactly do you define that physical mind, the one that was the object of the transfer of power?" *It isn't the physical mind; it is the material mind. Not even material mind—the mind OF MATTER. It's the mental substance contained in matter itself, in the cells. It's what was formerly called "the spirit of the form," when speaking of mummies whose bodies were kept intact as long as the spirit of the form lasted. It is that mind, that altogether material mind.* That was like a revelation. Suddenly, I saw a silent little girl gazing at a certain mummy in the Guimet Museum.... She had traveled all that way, fifty years of "descent," to get there.

To the place that survives death.

Once stabilized, the mind of the cells never stops. It keeps vibrating anywhere, even in death.

It is the site of the transformation of death.

That is where Mother was entering, more and more.

A dangerous ... unknown.

And time, too, changed: *There's no more time. It's as if another time had entered this one.*

In the mind of the cells lies the double secret of the transformation of death and the transformation of the body—perhaps one and the same secret.

TWELVE

# The Permeation

It was an increasingly strange life, incomprehensible for everyone, and Mother was wise enough not to say anything, like Sri Aurobindo, otherwise people would have believed her mad. And yet She spoke to me—though less and less, not because She had to remain silent, but because She was more and more eaten up by the invasion of people, their quarrels, their lies: *An onslaught of Falsehood*, She said. It was the year of Bangladesh. And She had to live in the midst of all that, every minute of the day in the midst of all that, while her body was undergoing a sort of slow hell—smile at everyone, answer everyone, accept everyone, swallow everything. Anybody else would have been exhausted, spent, completely drained by that incessant throng in her room. No, it was not exactly a peaceful retreat. And sometimes, more and more often, the precious hours—precious for everyone—that She set aside for me were nibbled away, eaten up by sordid squabbles: *They make my head spin horribly with their stories.... I would be covered by a crust as black as coal if I didn't constantly, constantly purify myself.* And when I arrived to see her, after waiting in line for an hour, two hours, She was so pale that I simply sat at her feet, in silence, and I did not have the heart to ask any questions. Many secrets have been lost, many links missed because one person did not like his next-door neighbor or another had betrayed someone else. It is heartbreaking. I looked at the "secretaries," and I felt like saying, "But don't you understand that these few moments with her are of interest to the world? ..." They would not have understood. That's how it is. And Mother was seeking the transformation of the world, so She really had to include each and every element in the crucible. But the crucible was pure poison. Yet, in the end, because of that impossible contradiction She was in, because of that increasing silence and the increasing invasion that seemed slowly and inexorably to push us apart, to cut me off from her, I began to touch something else: She was building another channel of communication between us, and from then on, those extraordinary moments of silence with her became the best "explanation" of the impossible path. *Shall I take you along?* She would say. And She would take my hand, and then ... it was not even a "trip," because you did not go anywhere "out there," but you began to feel concretely, to touch inexpressible, nameless secrets that were like the birth of a new being within yourself, a new perception, almost a new substance, a substance that was everything at once—being and (tremendous) power, vision, knowledge and love—as if, at that level, everything were made of love. And we moved with such a perfect accord within that substance that the least movement of one was perceived and followed by the other, and there was no longer any "other": *It was as if your body were embraced in the same cellular movement.* And today [1975] as two years ago, I feel the same experience continuing, as if She had built an everlasting channel of communication. And they say She is dead. Or they say She is in a "subtle body." But that body is more concrete and living and physical than all their ghostly gestures! If that body is "subtle," then subtleness is decidedly more solid than all their empty rumblings.

But the fact remains that She was eaten up.

And as always (this is the strange peculiarity of this path), it is as if the solution came out of the sheer contradiction, as if the difficulty, the obstacle, the impossibility were the key itself. No, She understood nothing of the path, but it came alive naturally under her footsteps, minute by minute. And if there was the slightest attempt to go against or "rectify" the difficulty, the key immediately went away. Mother accepted EVERYTHING. The destiny of the world and the sordid squabbles were one and the same thing, without any degree of priority. It was the *same* thing. It was the destiny of the world every minute of the time; everything was the destiny of the world. One was more and more reminded of Sri Aurobindo taking a detour so as not to disturb a sparrow's nest over his door. That detour and that bird and that work, forever incomplete because of thousands of epistolary detours of one person or another, were one single Movement. The work is at each instant. There is not a single detour. When we understand the absolute meaning of each instant and of each thing in that instant, we will be very close to innumerable life and to a body that goes everywhere at will.

**Iridescent Materiality**

But how was *that* body going to be infused into this one? That was the "burning" problem, if we may quite adequately say. How was that increasingly formidable vibration, that tidal wave of golden power going to modify the bodily substance, or even simply penetrate the body without destroying everything? The experience had in fact begun years before, at a time when nobody knew what it meant or where it led; as early as 1961 Mother had noted "the little speckling of vibrations that seems indispensable in order to penetrate matter," as if it could not penetrate directly without scorching everything. A fine multicolored speckling, which seemed to be the first "adaptation" of the supramental substance so it could infiltrate the body. She had seen the same thing some fifty years earlier, in 1906, but far away, at the end of a "breathtaking fall," and it is remarkable to note the accuracy of that experience; when She opened her eyes, at the end of the experience, She wrote, *There are also luminous, golden clouds, but dense, not transparent, and in those clouds are small dots of iridescent light. I felt it was the materiality brought by the trip.* The materiality of the new world. And it seems that through the years, imperceptibly, layer after layer, the fine iridescent speckling came closer to the pure body, pure matter, truly as if we were separated from our own body by layers of consciousness without being aware of it. The body is what is the most removed from us! Naturally, since all the atavistic residue is in the way, but the closer we get to it, the more we uncover older, primary layers—the residue of the animal, the vegetable, the mineral—as if the end of the journey were in fact the beginning of the world. We are millions of years away from our body! Yes, the very one we touch right here, which seems to stand so squarely on its two feet. But it is an enormous crust. That is what makes us die. And the whole phenomenon (since 1906, in fact) is the slow permeation or infiltration of that speckling through this dark covering, to the pure little cell, the direct body. Then what was all-the-way-out-there is all-right-here. When we feel it is all-the-way-out-there, we talk of "subtle" world and "subtle" body and "vision"—a dream.... And when it has come here... well, it's all physical!—a single physical separated by walls of consciousness and layers of filth. Within the pure little cell both worlds are ONE; the "subtle" is as perfectly physical as John Doe in jogging pants (maybe more so). It is exactly what slowly and patiently and carefully happened in Mother. And as that fine speckling penetrated her body, it is as if the "other world" simultaneously forced its way into this one, and She suddenly found herself, eyes wide open, in "Sri Aurobindo's residence," the so-called lining of this world where the living and the dead stroll together as if nothing had happened.

Then it becomes clear that what creates the veil between the two worlds is the physical mind, and what causes death is the physical mind—one and the same thing. A coating of false matter or fossilized matter over . . . an immortal *physical* reality. Death is not on the other side; *we* are on the mortal side, and we die precisely because we are not in the earth's immortal reality, in the true *physical*.

How to make that true physical pass into or infiltrate the old crust is our whole problem—truly a question of life or death for our species. One could say that, for the first time in the world, Sri Aurobindo and Mother raised the real question. For the first time someone found the real evolutionary question, which is neither escaping in little heavenly bodies nor being scientifically confined in some irrevocable, legal and mortal matter. Then, our millions of years of struggle and pain truly make sense. The earth becomes meaningful. Otherwise, it is simply a scientific monstrosity, or a similar religious monstrosity on the other side. We get out of this contradictory nonsense. We get out of the false dilemma that has obstructed the true paths of the work for so many centuries—but perhaps the time had not come. Even our science and religion have done useful work by bringing us to such a painful and absolute point of contradiction that the earth *had* to emerge into a third position. We are there. We are undergoing the *physical* transition to that third position. This is not a philosophical question; this is a question of flesh, bones and guts. The whole question is to know how the said flesh and guts will withstand that fine speckling of new light—oh, so old! Thus, Mother's body begins to take on all its reality as an evolutionary laboratory. Her endeavor is really OUR endeavor. And those who would try to divinize her or make her into a new religion are foolish. It is the earth that has to be divinized, it is humanity that has to be divinized, it is matter that has to be divinized. What we seek is the reality of the earth and of those who inhabit it—down with religions. Let us get to work on real matter.

## The Supramental Invasion

Yet, it took years for Mother to understand that that strange fine speckling, which She saw in everything, "associated with everything," even with her eyes open, was the very process of infiltration or "permeation," as She would later call it, of the true physical or real matter into her own body and the entire body of the earth—it was not special to her body, though her body, and Sri Aurobindo's body before, may have triggered the phenomenon on a world scale (before or simultaneously, since Mother had those experiences as early as 1906). But it was in 1962 that the phenomenon began to make a little sense for her—fifty-six years after the first experience! *It's as if the two [the true physical and the physical] were in the process of being fused more and more. Instead of one BECOMING the other, it's as if one were permeated by the other, and you can almost feel the two at the same time. That's one of the results of what is happening now. For instance, just a little concentration is enough to feel the two at the same time, which leads almost to a conviction that the real change in the physical is brought about through a sort of PENETRATION. The most material physical no longer has that sort of density which resists penetration; it is becoming porous and, being porous, it can be penetrated—in fact, several times I had the experience of one vibration quite naturally changing the quality of the other, the vibration of the subtle physical bringing about a sort of . . . almost a transformation, or at least an appreciable difference in the purely physical vibration. The moment I am quiet, the moment the body is still, I can feel those two vibrations, and the physical vibration becoming porous. That seems to be the process, or at least one of the most important processes. And it's not like something more subtle penetrating something less subtle without changing it; it's a penetration that changes the composition, that's what is important. It isn't just a greater degree of subtlety; it's a change in the inner composition. At the extreme, it is probably an action that must have atomic consequences. That way one understands ... (how shall I say it?) the practical feasibility of the transformation.*

An atomic alteration? Without exploding in the process?

And Mother added, *It's a very humble work in its appearance, very quiet. There are no illuminations filling you with joy and ... All that's all right for people seeking "spiritual joys"— that's the past. It is a very, very modest work. And at each step it's as if you had to be extra careful to prevent any disruption.* The "edge," as Mother called it, would become increasingly narrow and dangerous. *All the powers, all the realizations, all those things are ... a big show—a big spiritual show. You go from one fair to another to show your tricks. That's not it! Very modest, very modest, very unobstrusive, very humble, completely invisible on the surface. It takes years and years and years of silent, quiet, very meticulous work to get something visible, to get a tangible result....* "Each atom," said Sri Aurobindo. And the problem of time always came back.

The atomic level of the problem also came back. That very same year, 1962, Mother rather mysteriously remarked, *There's nothing to "change"! Look, as an analogy, consider what science found as the so-called composition of matter as far as the composition of the atom is concerned—there's nothing to change. There's nothing to change! The constituent element doesn't change; it's the relation that changes. Clearly, there is one and the same constituent element for everything, and everything lies in the relation. Well, it's exactly the same for the transformation.* And as I was left a bit perplexed, Mother clarified,

*You can only understand the meaning of the word "relation" if you take it in its scientific sense, in other words, your body, like my body, and this table and this rug are all made up of atoms, and those atoms are made of the same, SINGLE "something"; and it's only either the movement or the relation of that "something" which creates the apparent difference: different bodies, different forms. Well, I am saying that the Power must change that intra-atomic movement; then, instead of disintegrating, the substance will follow the movement of transformation. Do you understand? It's all the same thing, one and the same thing! But it's the relation in things that we must change. Then immortality is clearly within reach! It's just the fixity of things that causes them to disintegrate—and that's a purely apparent disintegration, since the essential element remains the same everywhere, in everything, in decay as well as in life.*

Was it merely an analogy that Mother used? Or should one take her literally? . . . Clearly, if one conceives that the supramental Power is the very power that animates and constitutes the atom, then it can operate what it has constituted— an impossible feat for us, except by the dreadful methods of the sorcerer's apprentice. But what would happen? . . . We lack experience, as Mother would say; we can only note the possibility and understand that that "change of relations" could only take place with extreme precautions—and time. *Actually,* She remarked in conclusion, *that's just the constructive Will. The constructive Will is eternal and infinite—it's obvious—therefore there is no reason that what was fashioned by it should not be part of that immortality and infinity; it isn't a necessity that things should disintegrate in their appearance in order to change form. That's not indispensable.*

Eight years later, in 1970 (one can appreciate how "very modest" and careful the process was), Mother remarked, *The union between the two, the subtle physical and the material physical, is taking place all the time—day and night, day and night. You could almost say that one is trying to replace the other. It's like a sort of . . . not exactly a fusion, but a permeation (it's truly a permeation), which doesn't push the other away, but . . . in the long run it will probably transform the other. That subtle physical is working to replace the other, but not through elimination—through transformation. And I can see (I see very well since I feel the two at the same time) what an enormous work this is. It removes the fixity. Our physical is brittle, and it removes that brittleness: what used to break now bends; what used to crumble is now fluid; it's becoming . . . It's very strange. Difficult to explain. But then, what a colossal work, you know. The experience is just now in progress. It began many, many months ago [years, in fact] in the more subtle, and little by little, very slowly, it is moving down to a more material level. . . .* The layers had all been worn down, and now the pure, direct body was reached. And simultaneously, the "other" world, the so-called subtle world, the so-called other side, came into this one (not "came," for it had always been on this side—it showed through). *Last night,* noted Mother in that same conversation of 1970, *it was truly remarkable. You could not have said, this is the subtle physical and that is the material physical; they were . . . surprisingly one within the other. You didn't have the feeling of TWO things.* The joining of the two sides was accomplished. But to tell the truth, that "other side" did not concern me in the least, even if it was on this side; the earth, the visible earth is what interested me, and I asked Mother, "But what is happening in terms of the *earth*? How is the permeation of that subtle physical working in the earth?" *But just like that!* Mother exclaimed. *Like that. That's what happening. That's how it works—permeation.* And I insisted, like Saint Thomas: "But is it happening on an earth scale?" *Yes.* "In each and everyone?" *Yes.*

And the experience grew, intensified: *I don't feel I am at all dreaming anymore, not at all. It has nothing to do with a dream;*

*it's a continuous activity. . . . The "over-there" continues here, and it is as real, as tangible as physical things. And there, people who have a body and those who no longer have a body are all mixed without its making any difference. They have the same reality, the same density and same conscious and independent existence. The physical seems less imperative, less . . . Before, you felt that, well, it wasn't exactly what people call a "dream," but it was a more subtle, less precise consciousness, while the physical consciousness remained altogether concrete and precise. But now that distinction is . . .* And Mother stopped, her gaze fixed in front of her, toward Sri Aurobindo's tomb. *It's as if that subtle physical world were trying to ENTER this one (there's indeed a great power in this), but I don't know how to explain it. It's as if it wanted to force its way into this world.* "An invasion by that subtle physical?" *Yes, that's what seems to be taking place.*

The supramental invasion.

But the more the two were joined, the more it came down and penetrated Mother's body, the more boiling and seething and hellish it became: the iridescent light turned into molten gold. Truly a small death every minute. And the same thing in the earth's body. Two years later, in 1972, the phenomenon had become even more tangible, striking, you could say: *There is a sort of golden Force, without any material consistency, yet which seems terribly heavy, pressing on matter. With the apparent result that it seems as if catastrophes were inevitable. But concurrent with that perception of inevitable catastrophes are solutions to the situation, events that appear themselves truly miraculous. It's as if the two extremes were becoming more extreme, the good becoming better and the bad becoming worse. Like that. With a colossal Power PRESSING on the world. And in circumstances themselves, many things that ordinarily happen indifferently are becoming acute—situations, differences are becoming acute; ill wills are becoming acute—and AT THE*

*SAME TIME, extraordinary miracles: people on the verge of dying who are saved, inextricable situations that suddenly find a solution. And the same thing is happening in people: those who know how to turn toward and sincerely call the Divine, who feel that there lies their only hope, who know it's the only way out and sincerely offer themselves, well ... within minutes it becomes marvelous—for tiny little things. There's no big and small, important and unimportant—everything is the same. The values change. It's as if the perception of the world were changing. This is to give a sort of idea of the change that the supramental descent will bring about in the world. Things that used to be indifferent are becoming really categorical: a small mistake becomes categorical in its consequences, and a little sincerity, a genuine little aspiration becomes miraculous in its outcome. The values are amplified, more precise. And the same for the body: the slightest thing seems to result in disproportionate consequences—good as well as bad. The usual "neutrality" of life is disappearing. And this is becoming truer day after day, hour after hour. The feeling that that Force IS effective, is REALLY effective—you see, it has the power to move matter: it can produce a material accident, and it can save you from a purely material accident; it can suppress the consequences of an absolutely material event. And it is stronger than matter. Things are no longer as they were. There's truly something new—things are NO LONGER as they were. It's truly a new world. We can call it "supramental" to avoid misunderstanding, because the minute you say "Divine," people think of God and that spoils everything. It isn't that. It isn't that; it's the descent of the supramental world, which is not a fiction. It's an absolutely material Power, which ... [and Mother smiled] doesn't need any material means! A world seeking to incarnate in this world.*

It was the year of Bangladesh, Watergate, the murder of the Israeli athletes at the Olympic Games in Munich, the withdrawal from Vietnam. Also Nixon's first trip to Beijing.

And the problem of that body's transformation is starting to be definitively linked to the world's transformation. It is one and the same burning "permeation" here and there. The miracle and the catastrophe side by side. The precipitousness of Death and the transformation of Death.

What is going to happen?

THIRTEEN

# Cellular Time

**Cellular Interdependence**

It was more and more "a race between transformation and death," as She had said twelve years earlier, in 1958. She received blow after blow. It was more and more "the onslaught of Falsehood," around her as well as in the world. *It's as if all the Falsehood were coming out because of the pressure. As if there were poison, you know, and the pressure removes the poison by making it come out—and does it come out!* . . . Clearly, it was not a degradation or a sudden depravity; what had been hidden—hidden for millennia—was just coming out in the open. And coming out! It had to come out of its hiding place in the animal sediment around our cells so the "other thing" could emerge—that was the very sign of the operation. The famous *panis* of the Vedic rishis, the wolves and devourers and gnomes of every kind hidden in the "cave" and preventing the "dawn," the blossoming of the "sun in the darkness," the cleaving of the "mountain" were all out in the open. They could no longer hide anywhere. It is the time when everything comes out into broad daylight. And comes out. . . . Things are unmasked, people are unmasked. It is not a "worse" time; it is the time of truth. And her body grew thinner by the day. *But you must know,* She wrote to one of the disciples, that in each of my children and everyone of them, whenever they speak or act under the impulse of falsehood, it acts on my body like a blow.[1] They may have realized it, but the blows continued nonetheless. Dressed in impeccable white linen, they would climb the small winding staircase covered with a golden wool carpet: *They feign good will. But the vibrations inside still belong to the world of Falsehood.* Oh, She was not in the least duped by her children, and She loved them, but . . . *This body is getting frightfully sensitive,* She remarked in early 1970. Of course, it was everywhere, in everything! And She scolded herself: to feel pain, to be capable of pain was still the sign of an "I" somewhere in the body, an "I" that felt. *At bottom, what still clings to the illusion of being something separate must disappear. It must say to itself: "That's none of my problem. I don't exist." That's the best possible attitude it can take. Then . . . it merges into the great universal Rhythm.* That movement of fusion was what one could feel taking place more and more in her body. Instead of shutting herself up in the pain like all other bodies, of stiffening and building another crust to protect herself, which in the end is a crust of death, She tried to shatter the last remnants of wall, the last little film that catches pain and accumulates death. Yes, but . . .

One day, a small incident showed the exact extent of the problem. The incident took place in the "music room," where sometimes Mother still used to receive people. *I had a very clear experience,* She related to me after the "fact." *I was with X., who was in a terrible state of excitement, rebellion, confusion—everything you can imagine—and who kept furiously throwing all that at me for close to three quarters of an hour. I was sitting there—and I didn't notice anything! I laughed, spoke, acted, moved, and the body felt per-fect-ly comfortable. Then I came back to my room. Y. and Z. had heard X.'s screams (he screamed like a madman). They had heard it all, and they were full of horrified pity for what X. had forced me to endure—and IMME-DIATELY the cells felt tired, a terrible tension they had not felt during the scene, not a single minute! When I got up to leave X., everything was charming—we were just playing* [just like Mother!]—*and the minute I entered this room, I was overcome by fatigue and tension . . . WHICH CAME FROM THEM! That gave me an interesting measure of the interdependence. The body follows the action very well, and everything it must do it does, but there are consciousnesses around that feel or think in a different way, and that still has a very profound effect. Mind you, the consciousness is not affected: it's very clear-sighted, constantly aware of the play and conscious of incoming forces—so how is it that, being conscious of the incoming forces, those forces are still capable of directly affecting the cells as they do? . . . It's a problem. It's the sign of a cellular interdependence that makes the program very, very difficult.* Indeed, the "program" was going to become more and more difficult.

Then Mother added the following, which showed another aspect of the problem: *So what is required is an all-powerful vibration that would squash all that, brrm! . . . But, as Sri Aurobindo said, if that came . . . it might destroy too many things?* This is the heart of the contradiction: on one hand, an all-powerful Force is needed to overcome the obstacle, and on the other hand, if that Force showed its face ever so slightly, it would mean the destruction of everything in the vicinity that resists. *And they were vibrations of good will: Y. and Z. had no hostility in them, absolutely none—the hostility was before, with X.! The rebellion, etc., did not have any effect.* The obstacle is not just in the "evil"; the formidable obstacle is also in the false, sanctified, sentimentalized, constitutionalized and medicalized "good." There were doctors around Mother. She was surrounded by all sorts of people who wished the very "best" for her. *After that, I said to myself, "How little we know! How little we understand of what actually IS—the working mechanism."*

She was seeking the mechanism.

How to overcome that cellular contagion... without crushing everything around her?

How to prevent the new species from being submerged again by the old one?

## The Central Experience

But there was one experience, repeated a thousand times during all those years since the first exit from the web of the physical mind in 1962, that gradually acquired a new depth, as if, under thousands of forms and faces, we were in fact always confronted by the same Experience, the same Reality gradually unveiling itself—actually the same matter, which we are all made of, gradually revealing what it is. But obviously, that real matter, as it is, can only reveal what it is by breaking a little the comfortable habits in which it has been shut up. And we scream, protest and suffer because we do not see that each thing, each accident, each circumstance is actually an immense and innumerable grace given us in order for us to arrive at the Secret, or at a layer nearer the Secret, and that everything is marvelously organized—but we always go counter to the Marvel, walking past it or laying one, ten, twenty coats of penicillin over it to "rectify" the abominable disorder or the abominable Falsehood. And we fail to see that that fracture or that crack was just the occasion for a new little reality to steal in between the debris or the shattered scales. If we knew how to apply this to the *slightest* thing, life would start teeming with miracles—for, in fact, the Miracle is constantly present; it is just occluded by a certain way of looking at things. Matter is a miracle. It is the last miracle. *The rejection of falsehood by the mind seeking utter truth is one of the chief causes why mind cannot attain to the settled, rounded and perfect truth,* said Sri

Aurobindo; not to escape falsehood is the effort of divine mind, but to seize the truth which has masked behind even the most grotesque and far-wandering error.[2] And Mother made the same discovery, IN HER BODY, with illnesses and death. *Instead of that egoistic reaction of saying: Ah, no, I don't want this. I want no part of it!—just let it come, accept it, and see what the solution is. In other words, instead of the age-old problem of rejecting life, rejecting the difficulty, rejecting the disorder and escaping into Nirvana, it's the acceptance of everything—and Victory. That, as far as I know, is truly the new thing brought by Sri Aurobindo. Not only the idea that it's possible, but that it's the real solution, and the idea that we can start doing it right now. I am not saying we'll complete the work now; I don't know, but the idea that now is the time to start, that the moment has come to start doing it, and that it's the only real solution. The other solution is no solution—it was certainly a necessary experience in the universal evolution, but escaping is not a solution—Victory is the solution. And the moment has come when we can try.*

The truth behind illnesses and death.

You need a lot of grit, that's all.

And you need to have faith in the real miracle of matter.

The simple model experience, which was soon to take on another depth, can be chosen by going back several years: *Suddenly, there's a sense of a sort of disorganization, like a current of disorganization, and the substance making up the body begins to feel it, then to experience the effects, and finally everything starts to get disorganized. This is the disorganization that prevents the cohesion of the cells necessary to make up an individual body. Then you know: That's the end. But then, the cells start aspiring; there's a sort of central consciousness of the body that starts aspiring intensely, with as much surrender as it is capable of: Your Will, Lord, Your Will, Your Will.... Then comes a sort of—it's not flashy or dazzling—it's a sort of... well, it's like a DENSIFICATION of that current of disorganization* [and this is where we are approaching the key], *and something STOPS. First comes peace, then a light, then Harmony—and the disorder has gone. And once the disorder is gone, there is an immediate sense, IN THE CELLS, of experiencing eternity, for eternity. In other words, a sort of change of time in the cells, a densification of time. The current stops. Well, that, exactly as I described it, with all the acuteness of concrete reality* [a heart attack or neuritis is something very concrete], *is what happens not only every day but several times a day. Sometimes it's very serious, meaning it comes massively, all at once; sometimes it's just something brushing by. But these are not flashy happenings; the person next to you doesn't even know about it. At most, he may notice a sort of halt in the outer activity, a concentration* [She would place the palms of her hands over her eyes for one or two minutes], *but that's about all. And you can't speak of this, you see, you can't write books about it. That's what the work is all about. It's a very humble work.* Then, as if She saw Sri Aurobindo's face and all those years of silence, those last, immobile years spent in the big green armchair staring at the Wall, She added, *Actually, proclamations, revelations, prophecies are all very comfortable; they give the impression of something "concrete." Now it's very humble—the feeling that it's very humble, invisible (it will be visible only in the results, a long, long time from now), not understood. And in fact, insofar as it is really new, it is not understandable.*

She herself did not understand. She groped her way through the forest. She had noticed the existence of a sort of "vertical time" which seemed to have strange properties, but it was such an elusive phenomenon, and, as Pavitra put it, it was an eternity that never had the time! Mother never had the time. She had to be positively on the verge of death to be allowed to breathe for a moment—one day, two days—then She would start again. But the picture kept unfolding, more and more precise and sharp; the blows fell on her right and left . . . so She would learn the mechanism. *It's as if the two extremes—a marvelous state and a general decay—were there together, one within the other. Everything, absolutely everything is getting disrupted—the people you counted on fail you, a kind of general dishonesty seems to be spreading, people are falling sick all the time* [in other words, their illness falls on Mother]. *As far as difficulties are concerned, there have never been so many of them, and combined: big difficulties together with ruinous ones. And simultaneously, for just a . . . flash (it comes for a few minutes, then goes away), a marvelous state (felt by the body), something unbelievable, you know, absolutely like the extreme opposite. As if it were trying to take the place of the other—but the other fights back furiously. So all the circumstances are like that, all the people are like that, from the government right down to the people here. And then comes that marvelous state. It comes for a few minutes in my body, then it goes away. That's the situation. That's what I've lived since . . . night and day, nonstop. Three minutes of splendor for twelve hours of misery.* Two states one inside the other, like life and death intertwined. And the constant cry of the cells in the midst of it—the call, the aspiration, the Mantra. And once in a while, the lived Marvel: *As if the air were changed into divine Presence. It seemed as if everything were touched, touched, penetrated* [Mother made a gesture of dots everywhere in the air], *but with . . . Most of all, there was a dazzling Light, a MASSIVE Peace, a Power, and then a sweetness . . . something . . . you felt it could have melted a rock. And this is the BODY'S experience, you understand, physical, material. The body's experience: everything, but everything is full, full, there's nothing but THAT, and we are just—everything is like something withered. The feeling that things are shriveled up, withered (not completely, just superficially), and that's why they can't feel it—that's why they can't feel Him. Because it's everything, everything. There's nothing BUT that, you see, you can't breathe without breathing Him; when you move, it's inside Him that you move. . . . Everything, but everything, the entire universe is inside Him, but MATERIALLY, physically. Physically. It's a cure to the "withering" that I am seeking. I feel this is fantastic, you understand?* A massiveness of Power on the other side of the web, beneath the withered sheath surrounding the cells. A different, miraculous air. Matter's miracle is right there, at every second—what is it that prevents us from living it? Or is this crust of old matter definitely incapable of letting it filter through? *And then, when I listen, "it" says things, too. I said to Him, "But then, why do we always look above?" And with the most extraordinary, fantastic humor: "Because they want Me to be very far from their consciousness!" So the body is trying to be fluid, to melt—it's trying, it knows what it is. It's trying—can't do it yet, obviously!* And Mother looked at her hands, at her bones showing through her skin.

But the experience grew clearer, more categorical each time, until the day She touched the key, after having had quite a hard time with the most ordinary of disorders: a dental abscess that puffed up her face. *I've had days when I really experienced every horror in the creation. They weren't at all moral things; they were mostly physical pain. Mostly physical pain. I had a ceaseless pain—a pain that doesn't let up, that goes on night and day—and all of a sudden, instead of being in that state of consciousness, you are in the state of consciousness of the exclusive divine Presence—the pain is gone! And it was physical, absolutely physical, with a physical reason for it: the doctors would say it was due to this, that, and the other thing, you know. . . . An absolutely material thing, absolutely physical—poof, gone! . . . You revert to the first consciousness—and it comes right back.* Then, in her slow little voice, Mother added, *But if you remain LONG ENOUGH in the true consciousness, not only does the pain vanish, but the appearance—what we call the physical "fact" itself—vanishes. . . . I have a feeling I've touched the central experience.*

Given time, the physical fact changes.

*But this is just a small beginning,* She continued. *One would feel in possession of the supreme Secret only if the physical were transformed. . . . According to the experience—the tiny experience of a detail—that's how the transformation works.*

Mother was slowly being led, "in detail," to that supreme Secret.

It was in 1968, one 23rd of November.

*But then,* Mother asked, *is that Consciousness going to be expressed in ONE body first, or is everything supposed to change together? . . . That I don't know.* Can *one* body undergo and live the experience without all other bodies? That is another question still. It may be the decisive question. But there is no physical impossibility. Two years later, in 1970, Mother made the following mysterious remark: *Last night, I suddenly saw a certain mechanism and I said to myself, "Oh, if only we knew that, how many things—how many fears, how many combinations, how many—would crumble, would become meaningless." It was so . . . What appears to us as the "laws of Nature," as "ineluctable" things, was absurd—an absurdity! It all crumbles with the true consciousness.* And Mother concluded, *Several times, people have told me they feel they are facing some ineluctable law ("there's this plus that, hence it's inevitable"), and my answer is always the same: If you wish it. It's YOU who decide it's ineluctable!*

But what was that "mechanism" which topples ineluctable laws, that something which forces the physical *fact* to change? It seems as if Mother had heard my silent question: *It's probably a . . . there's a POSITION to be changed, a position of consciousness that must be changed.*

A position of consciousness that cures the withering, allows the old matter to be infiltrated or permeated, and changes physical laws. . . . *I feel this is fantastic, you understand?*
Perhaps another position in time.
A densification.
Another time in time.
Not a time "out there," eternal, not a time that takes time—a time here and now, in matter.
A new time of matter.
Or its real time?

## The Change of Position or Massive Time

Clearly, if the rhythm of matter is altered, its entire structure must be altered. From the atom to our outer metabolisms, all of matter is built on or made up of a certain movement, and every scientific manipulation ultimately tries to act upon that movement, to speed it up or slow it down, or to overcome the forces and gravity it induces. Tiny electrons make up quite a solid wall with their movement. Everybody knows about the metabolic changes observed during the groundhog's winter sleep. And the caterpillar in its cocoon. Cryotherapy and induced low temperatures are still a science in infancy. Recent research on the drosophila (fruit fly) suggests that the cells that do not belong to the germinal line, those which neither reproduce nor regenerate themselves, and hence age, such as brain cells, for example, only age because of the "noise" accumulated from all the external stimuli, and that that "noise" could theoretically be erased, as one erases a magnetic tape, by cooling the said cells.* Cooling, that is, altering the rhythm. I am

---

* According to the work of Professor H. Atlan of the School of Medicine of the University of Paris.

no scientist, and I know really nothing of their theories and manipulations, but the fact remains that all alteration of time or movement causes an alteration in the structure, and that ties in with Mother's central experience: "If you remain long enough in the true consciousness, the physical fact changes." But scientists only know of superficial, external time, so to speak, which they must handle roughly and rather barbarically, like the sorcerer's apprentices that they are, without truly changing the structure of matter, except in catastrophic ways—they do not have the fundamental key. They do not have the key to the consciousness that produces movement and all the forms of matter resulting from it. They do not have the secret: Matter = Consciousness. One could say that Mother's whole experience is to arrive at the consciousness of matter, at the pure substratum or pure Movement, at the vibration around which every form has been generated and encrusted. There were the distant "golden clouds" She saw as early as 1906, the gold powdering of 1958. Then slowly, very slowly, that speckling—which seemed to take on different colors, to become iridescent, as it traversed the layers of old matter, the old evolutionary crusts—started to coagulate again, to become dense, homogeneous, once its course of infiltration was over and it reached the pure, direct, veil-free bodily substance—once Mother was able to withstand the "boiling porridge." It was "like molten gold," She said the first time. "It was very thick . . . with a surprising weight to it, you know." In other words, the "bottom" of matter had pierced the web and linked up with the surface of matter. And in 1969, when She spoke of that eternal same thing beginning to emerge, She called it "the new consciousness"—it was nothing "new," but it was perhaps the first time on earth that real matter, the real Movement, the pure Consciousness in the depths of matter, was emerging into the terrestrial field—and She said the following, which matches the pattern of thousands of little scattered experiences that came throughout the years as harbingers: *There's such a power! A formidable, absolutely formidable power. As if everything were swollen with power. An almost concrete power, I don't know . . . [Mother felt the air between her fingers], it's a light, but it's an almost tangible light: if it passes through your fingers, you feel it passing, so concrete is it. A deep-gold light. . . . One could say this: comparing the consciousness, not of ordinary humanity, but the consciousness of higher humanity to this consciousness, the feeling is that the moment the human consciousness sought to be in touch with higher things—to purify itself of lower movements, to widen itself—it became fluid, transparent, ethereal; whereas not only is this one far superior to the other in terms of vision and perception, but it is SOLID and concrete. There's a feeling—it's so strong!. . . . Solidity. In terms of consciousness, this has been the greatest change of my entire life, and I've had a lot of them and I've worked hard and—it was nothing compared to what has happened since the 1st of January (1969). To the point that the body feels like a new person. . . . But that's not enough.*

Solidity, massiveness, density—always the same pattern. Something that gets denser.

Thus, She was reaching the heart of the story. The "golden clouds" were right there. All the same, it was a slow hell that was set in motion in her bodily matter under the influence of that formidable agent—apparently the very agent of transformation, since it is the first agent of every material creation. But "that's not enough." A further adaptation or something else seemed to be necessary for the agent to work without smashing everything—and without taking centuries, which Mother could not afford in that old, diminished body. And this is where we approach the most enigmatic part of Mother's forest. For, while the phenomenon is taking place, it is incomprehensible, unintelligible, "something," one phenomenon among thousands. It does not have any logic; the logic will come later, once it is "understood." That logic is what I am trying to track down among the dense underbrush and miraculous thickets of the thirteen volumes of the *Agenda*.

We find a first inkling of the phenomenon back in 1961, but there are actually thousands of minuscule phenomena that look like nothing, show their face, disappear, only to reemerge five or six years later with their full meaning, as if they had progressed underground, making it impossible to know when things truly began, as if the transformation were really thousands of similar phenomena, without date and without specific "moments." *There are none of those sensational things that are interesting to tell,* She said to me that day in 1961. *It's a humble work of every minute. . . . Each day and constantly, night and day and at every moment, tiny, very tiny things happen—it isn't interesting. These are lines of experience, you see, which are followed—you would have to take notes every second—and at some point on the line, suddenly, you discover something. . . . For instance, there is that sort of mental-type activity (you can hardly call it mental) in matter* [the physical mind] *that starts up—it's disgusting. I still haven't been able to eliminate it completely. But there are times when it stops completely—oh, sometimes, when I walk* [while repeating the Mantra], *it's held like this, tight.* [Mother made a gesture as if She were suddenly frozen in place: the physical mind stops, resulting in the immediate opening of the web, an invasion of massive Power.] *The trouble is that for an ordinary consciousness (I am unfortunately surrounded by a lot of people with a very ordinary consciousness), it looks like a state of stupor, of stupidity, of coma, or torpor; it has all those appearances: something that becomes completely STILL, unresponsive, stopped dead. You can no longer think, no longer look at anything, no longer react, no longer do anything. But there's still everything that comes from the outside, things that try to interfere with and interrupt that state, but if I can prevent them, if I can KEEP THAT STATE, after a while it becomes something so incredibly MASSIVE, filled with such a concrete power, such a massive immobility, oh! ... It must lead somewhere.*

Indeed, it led somewhere—six years later.

Then She added, *But I wasn't able to keep it long enough (you've got to keep it for HOURS). I wasn't able to do it. There's always something interfering....* Six years later, the problem was still the same. There was always something interfering, not to mention the onslaught of people. *And when the body is abruptly pulled from it, it's as if it lost its balance. There are a few bad moments.*

It was the beginning of that "change of position" of the consciousness, which was to have the capacity to change the physical fact.

And once again it is remarkable to note that that first change of position of the consciousness coincided with the first attempt at piercing the physical mind. On the other side of the cage a different time begins, a different, *material* time, not a spiritual time (unless Spirit is one with matter). A massive time. Materialist and practical as I am (but Mother was no less practical than I!), I asked Mother, "But what happens when everything is still like that? Does something happen?" *Something happen? ... I don't know. But it is something IN ITSELF. When the body becomes conscious of that, it means it has emerged from its narrowness—it's the same Infinity one encounters when going outside of one's body. It's very difficult for the body to have it, very difficult: there's always something vibrating or stirring. For this is not just silence; it's immobility* [Mother made a massive, compact gesture, as if it could be cut with a knife]*, without tension, without ANY tension, effort or anything whatsoever. It's like a sort of eternity—in the body. It's as if it put everything back in order, yet nothing moves.*

An eternity of the cells.

And it took place while She walked, repeating the Mantra. It must mean there is a *true* immobility of matter behind its apparent movement, and those who would attempt some kind of cryotherapy of the transformation, who would operate on the appearances without operating on the essence, would commit a serious blunder—the same blunder as those who think they master atoms by smashing them in their cyclotrons, because consciousness is what needs to be operated on, the movement of consciousness is what needs to be changed in order really to change matter. Otherwise we merely touch the grimace of things, and we only create monsters and parodies of power. We can "cryogenate" John Smith all we want; he will only be a cryogenic John Doe, perfectly fossilized in his own filth.

When scientists become yogis, they will understand matter. And they will master it.

It is perhaps the time forthcoming.

The time of the physicists of consciousness.

The change of position on an earthly scale.

For, having the key to the consciousness of matter, the beings of the supramental species will indeed have the power to fashion matter directly, by emitting the appropriate vibration within their own matter. They will act from matter to matter as today we communicate and act from mental consciousness to mental consciousness. This is why Mother said, *For man, the supreme realization is comprehension—to comprehend things. For the Supramental, realization is Power—creative Will*. Like the beings on that supramental ship who shaped their bodily appearance or their means of transportation through the simple exercise of their will.

The position of consciousness must change and must shift from the mental level to the cellular level. We must move from reflected time to cellular time so we can operate on matter directly and change its structure.

There is a logic in that forest after all.

## The Contradiction

And Mother kept moving on.

The problem became more urgent every day or, rather, more defined every day—our problems are insoluble because we do not know what the problem is. It was no longer at all the problem of death as we conceive it or see it: an aging body, ailments piling up, deterioration, decay. She had seen, experienced the unreal absurdity of all those "natural laws" the moment She had gone through the web of the physical mind. That simply does not exist; death is something else. *I have seen more and more, more and more, that everything that happens, all the people we meet, everything that happens to us personally, is constantly putting us to the test: you pass the test or you don't. If you pass, you make a progress; if you don't, you must take it again. And now it has become that way FOR THE BODY: pains, disorders, threats of dislocation. . . . And there's constantly that consciousness, you know, straight as a sword, inside, which says, "You will hold on, won't you?" So you keep quiet, very quiet, and you call—you call the Lord. You repeat the Mantra, which comes automatically, and . . . peace is restored, and after a while the pain vanishes—all, absolutely all the threats vanish one after another. There are so many proofs—so glaring they are impossible to refute—of that marvelous Presence, so simple, so simple and so total, in everything that comes, in everything that happens, in the slightest detail, so it can lead us as quickly as possible to the transformation. There's this extraordinary experience that, the moment you take everything that comes to you as a means of learning to be what you should be, you immediately feel a marvelous, all-powerful and concrete Presence. Then you understand that nothing is impossible. . . . Just a drop of THAT.* We must learn to be. All the difficulties of evolution are there to teach us to be that which has the power to overcome the difficulty. On the other side of the web is the perpetual, or rather natural, miracle. There is no death *there*; at that level, at the pure cellular level where the great primordial Consciousness-Power-Substance, which Sri Aurobindo called the "Supramental," moves, death is impossible. It is real matter. So where exactly is death? Where does it hide? If *all* of the body-consciousness shifts into the right position, there is no longer any material reason to die. Is there something, in the body, that cannot or will not shift into the right position? A sort of unshakable point or recess of death through which we remain vulnerable? What? Where is real death, finally, if it is not in any of the pseudo illnesses, pseudo aging, pseudo wear and tear? What is it, in the body, that refuses *That*—life? And Mother described smaller and smaller concentric circles around a central point. It seemed as if the more open, clarified and illumined the cellular level grew, the more the invisible dross stood out which before used to blend into the general opacity. Until the day Mother put her finger on it: *The feeling that death is only an old habit now, that it's no longer a necessity. It's only because . . . well, because the body is unconscious enough to, not "desire"—that's not the appropriate word—but to feel a need for total rest. In other words, inertia. . . .* At the end of all the concentric circles one finds the mineral again. The rock, the original stone. *Once that is abolished, there's no disorder that can't be remedied, or at least (of course, the area of accidents has not been studied, but I am talking of the normal course of events) no wear and tear, no deterioration, no friction that can't be remedied. There's just that residue (which is very significant), that residue of unconsciousness seeking rest—what it calls rest is the state of inertia. That is, a refusal to manifest consciousness. That's all there is.*

Mother was reaching the bottom.

It was 1967.

And then . . .

*And then, there's that ENORMOUS collective suggestion pressing down. The form which accepts slow deterioration because of the ENORMOUS weight of the collective suggestion—a habit going back thousands of years: "It has always been that way, and that's it." The great argument. Actually, it isn't true.* This was the other side of the problem. Inertia and a desire for rest in the body; ill-will and suggestions of death around her. And the two made up the collective problem, which was going to become increasingly painful, almost insurmountable. *There are minutes when the body feels it has escaped the law of death, but it doesn't last. It's just a minute, a fleeting moment, and everything returns to the way it was. But the consciousness of the body is beginning to wonder why it is so. Why? Why? . . . And then, there are the people who come here with all their thoughts. Some of them come in, sit down in front of me and start thinking, "This is perhaps the last time I'll see her!" Things of that sort, you know. . . . The place is teeming with all sorts of concerns, from anxiety at the idea that such a thing may happen, all the way to . . . [Mother laughed] a haste for the end to come soon: "Free at last!" Free to do any stupidity I wish! . . . There is a swarm of desires for it to die—everywhere, they are everywhere! And the body has become very conscious; it's quite sensitive to what comes from people. . . . So all this falls on me by the truckload, and it makes things rather difficult. There are also other things that come—the good ones—but those are . . . they're very few and far between . . . perhaps once, twice in 24 hours: a pure light, all of a sudden. Something that . . . causes what you could call a minute of eternity. This is good. But it's rare. A little flame. . . . And that Presence. That Presence, that Presence . . . And these cells are like children, you know: when they feel that, everything, but everything disappears, except that Presence. For this body it's like . . . a sigh of relief. It doesn't generally complain when it's in pain—it calls. It calls, calls, calls. . . . And it knows full well that it's not at all necessary, that if it only knew ... how to get into stillness, to get into silence, that would be enough. As soon as it does it ... But I am not entirely sure that all these pains it feels everywhere, constantly, do not come from ... are not the effect of all the ill-will. They are everywhere on earth, you see.*

The difficulty of the world was right there.

In the face of that contradiction, inevitably something *had* to happen: either death or something else. And in fact, through the contradictions or because of them, Mother was led to the solution. The contradictions were but the key to the other door.

## A Cataleptic Trance?

But one morning in January 1967, just as She reached the bottom of the pit, that last (or first) layer of fundamental inertia, that residue of the stone, simultaneously, and almost necessarily, a solution emerged, then a first outline of experience. It truly seems as if there were never any "problem," but only moments when one reaches the very center of the difficulty, and once there, the solution is there, along with the problem. One merely has to journey to the absolute center of the contradiction.

There had never been any contradiction!

The "solution" was hardly satisfying, apparently, but the experiences that followed were definitely of a very new and different order. Hardly satisfying, but it forced itself upon Mother as an imperative order, "as if" Sri Aurobindo dictated the solution to her. Indeed, that morning Mother abruptly stopped in the middle of the conversation, and in a neutral voice, articulating each word, in a very imperative tone, She dictated to me the following note (I might add that just prior to that, Mother had had a series of very important physical experiences that could have been a beginning of transformation of the body but had all been interrupted in the middle by the "onslaught of people": *I am not accusing anything of having caused the state to go away. It went away because the body was still incapable of keeping it, that's all).* This is how the note read:

> *Because of the necessity of the transformation, it is possible that this body may enter a state of trance that appears to be cataleptic. . . . Above all, no doctors! This body must be left in peace. Nor should you rush to announce my death [Mother burst out laughing like a little girl very amused] and give the government the right to intervene. Be very careful to protect me from any deterioration that might come from the outside—infection, poisoning, etc—and be extremely patient. It may last days, maybe weeks, and maybe even more. You will have to wait patiently until I emerge naturally from that state after the work of transformation is done.*

I was a little stunned, and I asked Mother, "So it's something that's going to happen?"

*It looks like it. . . . Because I realized that this [the body] needed—needed time, that's all.* The disciples never left her any time! *Instantaneous things are miraculous things that don't have the power to last; they do not constitute a STATE—the vibratory state of something that lasts. So that order came. It was very imperative; it was an imperative necessity, which seems to indicate to me that it will happen. . . . You see, I became aware of all that needs to be changed in order for this body to be able to keep that state constantly, for it to be there all the time. That's when the order came.* "But that means a good deal of restraint on the part of the disciples," I remarked. *Yes. Nobody should say anything other than: Mother is in a state of trance, and that's all. If they are exposed to the idea in advance, they might be more reasonable?* . . . The disciples' reason is yet another marvel of the world to be uncovered.

A few days later, Mother explained further: *I rather clearly saw that the work depended on the ratio between two aspects: the aspect of individual transformation (that is, the transformation of this body) and that of the general, collective and impersonal work. If a certain balance can be maintained between the two, we can do without that prolonged state of trance, but what could be achieved within a few weeks or months (I don't know) would then extend over years and years and years. Therefore, it's a question of patience—patience is not what is lacking* [but it started to be in short supply in the disciples]. *But it's not just a question of patience; it's a question of proportion: there has to be a certain balance between the two, between the external pressure of the collective work and the pressure of transformation on the body. If the instrument [the body] is constantly and infallibly able to do exactly what is expected of it, then the trance would not be necessary. It's only if there's some resistance in the execution.*

There was going to be a lot of resistance, more and more.

But Mother was not any more satisfied with the "solution" than I was (I have to admit it). It looked as if it were merely the foresight of a Consciousness that considers all possible accidents. *That possibility of transforming the body while in trance was announced to the body about . . . yes, some sixty years ago, and periodically since. And it was always accompanied by this prayer: May it be unnecessary, it's a method of laziness.* Mother was someone who always stood on her two feet. Generally She was sitting in her armchair when I met her, but strangely, to me, She always seemed to be standing and on the move. *It's a method of inertia*, She concluded. *The consciousness is more and more alert—but also alert to the point of being receptive to the possibilities of unconscious resistance. . . . Everything depends on the plasticity, on the receptivity.* "But what would be the actual use of that trance?" I insisted. *To transform what is not receptive. There are billions of elements in a body! So there's a whole mixture of receptivity and nonreceptivity. It's still very mixed. And this mixture is the reason that the appearance remains as it is. So it's an enormous task, you see. If it had to be done in detail, it would be impossible, but through the pressure of the force, it can be done. . . . Provided the pressure does not cause everything to boil and explode. So the trance would be a means of speeding up (relatively speaking) the process. But the work IS BEING done. The only thing is, it can extend over hundreds of years, you see! That's what Sri Aurobindo said: one must first establish a state of consciousness in which the collective life of the cells can be maintained as long as necessary.* Actually, that was what was happening with the development of the mind of the cells, the little golden vibration that kept repeating and repeating itself. But hundreds of years necessitate an entourage that follows and accepts.

Furthermore, when the hour is already so stifling, all those centuries make our blood run cold.

*It has always been said that the external form will be the last to change,* She continued; *that all the inner, organic functioning would change before the external form, the appearance (it's just an appearance, you know); that the appearance would be the last to change.* That is what the whole problem boiled down to, to that external crust, that "piece of bark," as She said, because inside there was a perfectly good cellular, *physical* body, independent, capable of traveling everywhere in the world and readily moving about on the other side of our veil. That piece of bark was the test, the bridge between the two sides—the apparently insoluble mystery, except through interminable centuries. *This appearance is, it seems to me, the legacy of primordial habits, habits of matter. You see, matter evolves from total unconsciousness, and, through the ages and all the ways of being, it returns to total consciousness—it goes from one extreme to another. Well, those habits of static immobility are what creates the need for trance. It should not be necessary. The need for immobility and immobile rest should be replaced by . . .*

By what?

A new state of matter was needed that would combine immobility and movement, the repose of death and the movement of life, the rock's inertia and the dynamism of existence. The point was to change the base of inertia that draws us all into death. A different time was needed in order to live in matter, a time that would no longer be a time of fatigue and decomposition. Nor a time taking centuries. Time is the very sign of death. To change time is necessarily, automatically to change matter's very foundation, its . . . encrusted "way of being."

A cellular time that changes the conditions of matter.

It was exactly the experience that was beginning to unfold, as if emerging from the very heart of the problem. She had to reach the heart. We must always reach the bottom, and only then does the door open.

And we remember Sri Aurobindo:

> *A voice cried, "Go where none has gone!*
> *Dig deeper, deeper yet*
> *Till thou reach the grim foundation stone*
> *And knock at the keyless gate."*[3]

In the depths of death is concealed the last door.

## The Innumerable Present

Exactly one month after Mother dictated that note on the possibility of a cataleptic trance, and just as She reached that ultimate layer of inertia, that residue of the mineral in the cells, that real bedrock of death, came a very little experience, which Mother did not fully understand at first, but which was the exact continuation of the experience of 1961, six years earlier, when She was seized by that strange "massive immobility." "It has to lead somewhere," She said. Things take a long time to evolve. At first, She thought She was *learning a way to rest without leaving the body,* and perhaps it was that, too, but it was mainly something else. For years—maybe eighty years— Mother had been in the habit of leaving her body whenever She wanted to rest. Within a split second She went all the way up into the supreme regions of consciousness, into that great still, white light (as all accomplished yogis methodically do, and as we all, or nearly all, do for a few seconds at night when we lose all contact, without realizing it). For humans, this is tantamount to fainting. There are no connecting links. We have not built the intermediary steps of consciousness, and when we "fall" into that, it feels like falling into a hole—from which we come back refreshed. This is how Mother was able to carry on her exhausting life for more than half a century while taking only two hours of rest at night. She went there as if She were completely at home. But as She advanced in the yoga of matter, She realized that that method, that sort of negation or oblivion of matter singularly complicated her cellular work. It was certainly a great relaxation and rest to be able to escape the clutches of the physical mind, but it was an escape above, not below, and meanwhile the cells and matter sank back into their primordial Inertia—a sort of death in white, we might say. In other words, just the opposite of the goal. It is exactly what preparing for death is—a refreshing death. That very convenient method was going to be radically and abruptly taken away from her, so much so that She *could no longer* escape her body. She would be bound there, with all sorts of excruciating pains, so She could find the solution *there.* No more switching off the current, no more numbing of the nervous sensitiveness by a little inner concentration. All the yogic tricks were gone. And they were really tricks, a super-medicine we wish all of humanity could have, but which only fixed the problem temporarily, exactly as others fix it with their sleeping pills and tranquilizers. It is really matter that must find its own solution—She was driven there by force.

*I was sure I was what people call "awake"; it had nothing to do with sleep. There was just the consciousness watching. But interiorized. And then, the will to get up at 4:30* [when Mother got up every morning]. *I looked at the clock once in the meantime—it was 3:15—and I was surprised and said to myself, "How come? It was 2:30 just a minute ago! . . ." Then it was 4:30: "How come? I just saw it was 3:15!" I was completely flabbergasted, because I HAD NOT LEFT my body. I know I had not slept, and the consciousness was completely still, with no motion to speak of. It had been as if instantaneous. And from time to time it happens to me also during the day. I enter a certain state (it lasts just a minute or two), a strange state: you are completely awake and completely conscious, and yet completely unconscious of the time and things around you—not exactly of the things around you, but NOT IN THE SAME WAY. I don't know how to explain it.*

The change of position was taking place.

The beginning of cellular time.

Actually, the beginning of a completely new state which could well introduce a new era of matter. But words are big, and when it happens, you do not know it is a new era: it is 10 o'clock in the morning, and it looks so simple, almost ordinary. Furthermore, we would be completely mistaken to think of it as a sort of definitive state, like the state of the fish or the monkey or like the mental state; it is just the *beginning* of something. It has to lead somewhere, as Mother said—we do not know when this "somewhere" will show up or whether its form will not change on the way. It is something being born, a new state of matter being born. Nobody in his right mind would think that the new species will be a sort of sleeping supergroundhog—and first of all, this has *nothing to do with sleep.* It is just . . . something that was going to develop in a strange but very rapid way. Perhaps we will only comprehend the phenomenon fully in twenty or fifty years, or a century, once it assumes its true form or attains its true expression in evolution.

One year later, in 1968, the line of the phenomenon had already grown clearer: *I am becoming lazy. . . . It's strange, it just FORCES ITSELF UPON ME: I follow a movement, and . . . I drift into trance. And it happens at any time of the day. I eat, and in the middle of eating, something comes; I follow the movement, and I remain rapt—then I look up and see everybody waiting!* Indeed, Mother would sometimes remain 45 minutes with her spoon in midair. "I also noticed it," I said to Mother. "For the last few months or so, I even had the impression that 'Mother is withdrawing,'" *No. I am INSIDE, far deeper inside than before—not inside here, inside "myself," but inside all things. . . . Extremely sensitive to all the movements of those around me.* And She added, *The relationship with external things is no longer the same.*

A whole new world was evolving, or a whole new way of being in the world.

"But what do you *do?*" I asked, because I was forever stuck on the practical side of things. *When I go off like that, "inside," I always feel as if I were . . . molding vibrations,* She replied. *And later (the next day or in the course of the day) I learn that something happened to someone—someone called me and asked me that. It's always a call. And that's an answer. Something happened to somebody, something became twisted. So I work on it, putting it straight again, bringing back the light, the right vibration. And that same day or the following, I receive a note: "I was in great pain" or "I called you." That's what happens.* But that "molding" of vibrations by Mother is precisely the beginning of a new handling of matter. This is ultimately how the supramental beings will mold all matter at will. It is a *state* in which matter can be molded. And if we regard those twisted little ailments or little cancers that Mother "put straight again" as negligible, we miss a very crucial point. These are not healing stories; it is matter getting straightened through the emission of material vibrations, an action of matter on matter. Direct transmission, without cyclotron, without sleight of hand, and without aspirin. *Something strange is happening which I don't understand, and it's becoming more and more acute: I spent more than an hour taking my breakfast, and I thought I had finished it in twenty minutes! I've lost all notion of time. I was certain I had finished it in twenty minutes, and I spent more than an hour—to eat nothing. I take a mouthful or a sip, and 10 or 20 minutes go by—I don't know where or how. I have the feeling I am in a light, but a light that . . .* "But is that light active, or what?" I insisted. *Yes. Oh yes, it does LOTS of things. But not—not that way. It's . . .* And Mother could not explain.

*It's the quality of time that is changing,* She said toward the end of 1968. *There's a sort of intensity of consciousness that alters the value of time. I don't know how to say it. It's just the beginning. . . . You enter a state in which time no longer has the same reality. It's something else. Very peculiar. It's an innumerable present.* What is most remarkable is that this concerns the *body.* It is the body that goes into that other state of time; it is not the consciousness soaring into ethereal regions where time ceases to exist, the old yogic trick. It is matter itself altering its own sense of time. But what does it mean for matter to change times? What *is* the time of matter?!

It is the time of its own trepidation.

And this is where the phenomenon begins to take on other proportions.

"It's just the beginning...."

*<br>* *

We are walking in a forest where nobody has ever walked before. In fact, we are now leaving the paths of universal evolution, which seem sufficiently charted with the discovery, implementation and operation of the mind of the cells—which will automatically work out its own results and achieve its own evolutionary formula in a new species. It will evolve its own way of handling matter, as mental man of the Neolithic Age evolved his. This book could end here, even on an unfinished note, for indeed evolution is always unfinished. It is a matter of time—and perhaps human collaboration. But do we have the time? This is the question.

We are now entering Mother's last mystery, which seems to hinge on that question of time, and the only way to tackle it is to follow the experiences in chronological sequence without trying to organize them in groups of the same nature (groups of what? nature of what?). In what way is this last mystery connected to our evolution? If we knew it exactly, it would mean a great rip in the veil. A Mystery that would perhaps elucidate all other mysteries and unravel all our immediate impossibilities. We walk forever beside a little "something" that alters all the data. That little something may be Grace ... or the next world's gracious smile—as if it were not already here, watching us and silently preparing its smiling coups d'état.

Or perhaps its worldwide coup d'état.

*Part Two*

A DANGEROUS ... UNKNOWN

FOURTEEN

# The Residue

## The Missing Side of the Atom

Secretly, the whole task was to illumine and transform the last mineral layer, that "inexorable rock in the depths" Sri Aurobindo spoke of, that primal tomb in the body.

All the time, we think we do things according to a certain perspective, and we construct all sorts of justifications and theories around it, then we uncover another layer of significance, and the theories are overturned—in fact, we simply journey toward . . . "something." Mother never made any theories; She walked, and She told what She saw as She went along, that's all. The significance was thus never distorted, no more than a river is. In a century or two, we will still be able to bathe in that *Agenda* and its water will be as clear as ever. We put up a nice little magic sign: "Cellular Time," but what is there behind it? We may be still very far from knowing it. It may be analogous to the time of the man who joined the course of some stars and built a whole world—but these are different stars and this is a new world. And those stars prepared these, which are preparing what? We are not certain that "cellular time" has all the meaning we are trying to read into it.

This is how it gradually appeared to Mother (what is quite remarkable is that, because She forgot everything, each experience was always new to her, lived for the very first time; the world was new at each instant, as fresh as if it were just born—pure, without oppressing scars from the past ... except in the cells, precisely, which were rediscovering their past, their very old past, the old mystery): *For the body, it's a transition from the inertia-induced peace—the calm that long ago was the result of Inertia—to the calm of Absolute Power. It's a difficult transition. It's difficult for the cells. It is that work of transition which is now taking place in detail, and it isn't easy.* It meant changing the very movement of the cells. It is actually what we might feel if our body were suddenly plunged into "death." It is such a radical change of time, such a different state, that it feels like death for everything else. Can one possibly understand? ... Mother learned step by step, without quite knowing where She was going, whether it was death for good or something else. Have we ever seen a living species evolve into another species ... while taking notes on the transition? That is a little what was happening. Mother's sangfroid is something that almost terrified me. I say "terrify" knowingly, because I saw and felt; She drew me a little into the "movement" of her cells and, well ... One must either enjoy going through that, with Sri Aurobindo's great irony or Mother's humor, or desperately yearn for something else. Then, if things break, it does not matter. In 1969 the picture was beginning to grow a little clearer (but the picture of *what,* no one knows for sure). *A great passivity is required for the Force to pass swiftly and reach the body,* She noted. *I see it each time there's a pressure to act upon one part of the body or another. It starts out by an absolute passivity, which is ... a perfection of Inertia—the perfection of that for which Inertia is an imperfect representation.* And Mother looked at me out of the corner of her eye. *It's particularly difficult for those who are very developed mentally. It's very diffi-*

cult, because all its life the whole body has tried to be receptive to the mind, to obey, to be passive, etc., and this is precisely what has to be abolished. All of a sudden, I understood the extraordinary miracle of Sri Aurobindo lending the cells of his fingers to his typewriter keyboard while his mind was completely still. And the current went through directly. The current that perfectly guides the little birds and the great stars.

*How can I explain?* Mother continued. *Mental development implies a constant and general awakening of the whole being, even the most material one—an awakening that also creates a state that's the opposite of sleep. While, on the contrary, in order to receive the supreme Force, the equivalent of immobility is needed—the immobility of sleep, but an ABSOLUTELY CONSCIOUS SLEEP. "Immobility"—I don't know how to say it. . . . But it's almost the opposite of Inertia in immobility. And that is how I now understand why creation began with Inertia.* Perhaps because only the original Inertia was capable of withstanding the overwhelming current in its entire purity? A noninertia, that is, a resistance, would have made everything explode (?) *So the whole task,* Mother continued, *is to find that original state again after having gone through all the states of consciousness—a whole course. . . . And one gets the feeling that only the body—receptive, open, at least partially transformed— is capable of understanding the creation, what we call "creation," the why and how of it (both things). And it isn't something thought out or something felt; it's something experienced, and that's the only way to know.* All the same, I did not understand how that "immobility" worked at the human level, and I stressed to Mother that, in the highest states known to me, there was an intense aspiration, even in the body, and this was quite the opposite of immobility. It was an active aspiration. So? "What is one supposed to do?" I asked her. "Should one let everything go slack or should one maintain that active aspiration?" *It's difficult to say,* She replied, *because I am convinced that each per-*

*son has his own path. But, for this body, the path is to have that active aspiration.* "But then the immobility is gone!" *It has succeeded in doing it. It has found the way.* "You mean both together? The union of the two?" *Yes, they are together.... And this is where the mystery of supramental time, or cellular time, begins to come to light. That's what the body has succeeded in having: a complete immobility together with an intense aspiration. And it's when the immobility is left without aspiration [i.e., pure Inertia] that it falls into a dreadful anguish, which "arouses" it instantly. That's it, you see, an intense aspiration. But it's absolutely still within, as if all the cells had become immobile* [the trepidation in the cells comes to a standstill]. *That must be it: what we call intense aspiration must be the supramental vibration. I have often said that to myself. But if, even for five minutes, the body slips into the state of inertia—of immobility without aspiration—it is awakened by an anguish as if it were on the verge of dying! It's as acute as that, you know.* [as acute as the point that separates death from immortality, which look almost like twin sisters with just a tiny difference of aspiration]. *And for the body, immobility is ... yes, it feels that the highest vibration, the vibration of the true Consciousness, is so intense that ... it's equivalent to the immobility of inertia. That intensity is so great that, for us, it's equivalent to inertia. That's what is taking place in the body. And that's what made me understand (because now the body understands), what made it understand the process of creation. One could almost say that it began with a state of perfection, but an unconscious perfection, and it has to evolve from that state of unconscious perfection to a state of conscious perfection. And everything in between is imperfection.*

A new time, which is a very ancient time, but conscious. The protoplasm recollecting what set it in motion. Only the body can understand.

And Mother's sketch indeed connects that of the original creation with the apparent immobility of the rock in a lightning-fast atomic movement: matter's very passivity is what enables it to withstand the staggering Energy, while providing it at the same time with an automatism very much akin to infallibility. At the end of the course, Mother found that original "perfection" again, with a difference that distinguishes the path of life from the path of death. And one cannot even say that the difference lies in the intense aspiration of the end product (Mother), while the beginning product is bereft of aspiration and consciousness. There is consciousness in the depths of the atom, there is aspiration in the depths of the atom, as in the depths of any form of movement—the very movement is the sign of that aspiration; it is that aspiration in the depths which caused the ascent of evolution in the first place. But the moment a "form" came into existence, even the form of an atom, there also came a confinement of the Current, "my" aspiration, as it were, like a child hugging her doll to her heart—it's mine—and a will or an impulse to keep that particle of "self" to oneself—the very image of inertia, the imitation or caricature of immortality. That is the beginning of death. Something appropriated that particle of current and will not give it up. So each time, the form must be broken in order to evolve to a higher form, or a higher aspiration. That inertia is the stability of each form, a coagulated, stereotyped movement that seeks to keep on being perpetually what it is. And this is why each form says no, no, no. Each form is a stupendous NO fearful of losing its life, of letting it leak out through the least pore, the slightest crack, and so it turns and whirls to build a wall of electrons, or something else, around itself and safeguard its egocentric gravitation. Inertia is the NO. Death simply hangs on a no. Yet that *same* Movement can make life endless . . . provided we let it flow, provided the "something" that says no finds the same rest, the same security and the same stability in an open immensity instead of a closed point. There is really no death; there is a difference of attitude. But it is the attitude of *matter* that must change. We must find the YES of matter. We must move from the inertia of a dead aspiration, buried in an individual vortex that gives it the appearance of an opaque tomb of eternity, to a living, open aspiration, bursting into a universal movement, so swift that it retains all the density of the walls of electrons without their hard opacity, and all the true stillness of eternity without the tomb. The Supramental, the cellular time, is precisely that which affords the missing side of the atom, the one desperately sought by all of evolution: immobility in ceaseless motion. An immobility one does not die of. The entire evolutionary course really amounts to finding again the all-encompassing immensity in a form that is not a tomb.

And one wonders if those scientists who are "peeling" matter layer after layer, constantly uncovering smaller and, it seems, more "massive" particles in the depths of the nucleus,* are not seeking a fundamental unity that is everywhere here—dense, without gaps, golden, in a time too swift for all their microscopes, a time in which the movement's ultimate velocity blends with eternity's immobile instantaneousness. The missing side of the atom. The one that the body may find before them.

The whole question is to know whether matter is capable of relaxing, we might say, without breaking to pieces, of open-

---

* A recently discovered *psi* particle is forcing scientists to consider whether the number of fundamental particles of matter is not 4, or 6, or even 18, thus getting ever farther away from the simple unity they are searching for. But they may be in search of a scientific myth, for if the ultimate unit of matter is indivisible, it has by definition no surface—and how could they measure that which has no surface? That "no surface" unit is exactly the omnipresent Supramental, without gap or division: the missing side of the atom. Matter's fundamental unit.

ing itself to the formidable Movement without resisting or holding back—which means one must be utterly immense, and yet be in a body, a form. This was Mother's whole cellular training for so many decades. This is the change of time or change of position which must ultimately transform matter, this hard piece of bark, congealed because it does not let the Current through and falls asleep in the false peace of death. Matter must find its own eternity. Then it will no longer seek to die, no longer contract in a position of death. That was the experience beginning to happen in Mother's body. *Looking at the days in sequence, this is what the body experiences: In a certain way, at certain moments, it is in the consciousness of Immortality; then, through influence (and from time to time through old habit), it falls back into the consciousness of mortality, and that's really . . . The moment it falls back into the consciousness of mortality it is seized by a dreadful anguish, and it's only when it can get out of it, when it can find the true consciousness again that that goes away. So it's either this or that* [Mother made a to-and-fro gesture from one consciousness to the other]. *And the other state, the state of immortality, is immutably peaceful, quiet, with . . . something resembling lightning-swift waves, so swift they seem to be motionless. That's how it is: nothing moves, apparently, in a formidable Movement.*

A cellular eternity in headlong movement.

"A moving immobility," Mother said.

Then, suddenly, Mother would close her eyes and go off into that "light which does so many things," her spoon in midair: an innumerable present, "a 'within-ness' in all things"—not in the least suitable for the life surrounding her, which watched her every move, ceaselessly trying to get something out of that body, to make it do one thing or another, sign this paper or that one—a thousand things for the most part full of deceit. Even to make her take drugs to cure her of her bizarre eternity, which some (many, in fact) viewed as senility. *The body is starting to*

## 166  THE MUTATION OF DEATH

*wonder . . . to wonder how it will relate to things, what will be the relationship of the new consciousness to the old consciousness of those who will still be humans.* This was in early 1970. Exactly seventeen years earlier, She had asked the same question of the intractable little representative samples on the playground: *Is it possible for one body to change without something also changing in the surroundings? What will be your relationship with others if you change that much? . . . It would seem that a whole range of things would also have to change, at least to some relative degree, in order for that body to exist, to continue to exist.*[1] The question was going to pose itself more and more painfully—who followed? It was not only in one body that She had to overcome the NO, but in all the bodies around her: *There is still a background (that's the main trouble), a background of unconscious negation that's behind absolutely everything. It pervades everything: you eat, breathe and receive that negation. . . .A colossal work is still required for everything to be transformed.*

And we seem to hear Sri Aurobindo:

> *The stubborn mute rejection in Life's depths,*
> *The ignorant No in the origin of things.*[2]

What was going to happen?
Where did that cellular eternity lead? To what impossible state among men? Was She going to be able to dissolve that negation, that residue of the original Inertia? And could one change a parcel of matter without changing all of matter?

Or else did that lead to another, deeper layer of significance where the YES—instead of confronting the NO and immortal life confronting Death—would change, together with the NO, into something else, a third state?

## A Question of Patience

Mother did not know. She knew cruelly nothing. *Why, but why am I not told what is going to happen? I don't know. . . . It's to force me to be in a very passive state, I think.* Perhaps because we do not know what we are supposed to fight against. We call it death, but what is behind it? There may not be any "against," but only successive steps, successive means to arrive at that "something," that ultimate mystery, and everything is just a means of fanning the fire of aspiration that will open the last door.

For now, death was that fixity of matter which led to the slow disintegration of the form. And yet the experiences kept multiplying, obvious, convincing, always following the same pattern: *The body is starting to feel in a very clear and precise way that the moment it feels itself—feels itself and feels the rest in relation to itself [i.e. the old egocentric position]—it falls into a hole; and the moment it feels the Force acting, the Consciousness acting, then this [the body's fixity] is only very relatively real. And the body is learning well; it sees that in every little detail, constantly: the moment it feels itself as "something" and the Force as "something else," there's a pain here, a pain there, this breaks down, that starts acting up—a complex and very nasty world. But when it makes a certain movement . . . (what shall I say?), the opposite of condensation, rather like a dilation, something like a dilation in the consciousness, then the limits tend to fade away, to vanish, everything becomes supple, and the pain disappears—PHYSICALLY. It's something it experiences day after day, sometimes in one place, sometimes in another, sometimes one thing, sometimes another. The body feels that matter's fixity is an illusion and that it can . . . be overcome.*

And one wonders if that "dilation of consciousness," that sort of acceleration of consciousness, which also eternalizes and universalizes, is not a more powerful movement than the one that congeals and freezes matter in its mortal position—unless old matter proves to be totally incapable of withstanding the change of movement without dissolving? Mother did not know. She felt that movement of fusion, but... *I have the strange feeling that it's like a shell, or pieces of tree bark, like a tortoise shell melting away. Matter, as it appears to man, is like something calcified that must fall away because it isn't receptive. But this body [Mother pointed to her own body] is trying.... It's trying— oh, what a strange sensation! A strange sensation. If one could last long enough for all that to melt away, then it would be the real beginning.*

To last long enough.

Always the problem of time. That is probably what was trying to establish itself in her body through that ever faster change of position, that sort of eternity in the cells. *In the true position, there is no friction.... The body itself feels that it must learn to live in eternity,* She said as early as 1962. Yes, but would it be given time to live in eternity?

Or else the cataleptic trance?

And the question became increasingly pressing: *There's an impression of fluidity, of plasticity, that stands out more and more as the true consciousness grows. The hardening seems to be the result of unconsciousness. The lack of fluidity, of plasticity, seems to be the result of unconsciousness. Such is the impression, not only in the body but for everything. With the growth of the normal state of consciousness there come a suppleness and a fluidity that completely change the nature of the material substance, and the resistance results only from the degree of unconsciousness, is proportional to the degree of unconsciousness. The apparent materiality lies in the degree of unconsciousness....* [This is indeed quite interesting. The world we experience, the matter we experience are fixed and opaque only through their degree of unconsciousness—and through nothing else. A kind of real illusion. An opaque illusion.] *And what is interesting in respect to this body is that I have the growing feeling of a still-unconscious residue. There are still layers that remain as a residue of everything that came before—the mineral, vegetable, animal and all that. So the part of the cells that's fully conscious is fully illumined, but . . . It's easy to see, in fact* [Mother touched the skin of her hand, visibly unchanged]. *The density doesn't seem to be the same anymore, but the appearance is quite the same. Those who have an inner vision see something, but it's only because they have the capacity of inner vision. The new way of being would be visible only to someone who has himself or herself the supramental vision.* And yet, that so-called inner vision or supramental vision is a material vision, since Mother saw material happenings: *I MATERIALLY see all sorts of things, but they are not visible to others. Yet I see materially. A strange state. . . .* We are brought back, always, to those two worlds of *matter* one within the other: an illusory, unconscious, opaque crust, and the other. And I looked at Mother. I could well understand the growth of that new consciousness, new vision, new body—but the rest, the "residue"? There lay the mystery, the bridge. Unless the illusion is shattered and all the old bodies in the world crumble to dust? That would make a lot of dust.

Mother seemed to have heard my question: *What I still don't know, what isn't very clear, is the fate of that residue. . . . In people's ordinary view, it's what they call "death," which means that the cells that couldn't enter that state of plastic consciousness are discarded. But the way in which the work is being done doesn't make any categorical distinction between conscious and unconscious groups of cells; there are just almost imperceptible states of variation between the different parts of the being. Hence the questions: where, what, when, how, what will happen? . . . It's becoming more and more of a problem. You see, the impression is that there's refuse, but the refuse isn't some-*

*thing discarded; it's something that vacillates, that lags behind, that makes an effort and tries—and that is quite willing. For instance, whenever there is a minute disorder in some part or other, or a pain, it no longer starts to fret and worry and want drugs or doctors, not at all; it calls . . . it goes, "O Lord." That's all. And it waits. And generally, within a few seconds the pain goes away. What complicates matters is the INTRUSION of outside suggestions, of thoughts, of ignorance, of impressions—the whole swarm of impressions around. Most of the time, it has no effect, but sometimes it creates a jolt. And that complicates matters a little. . . . Take, for example, the fact that I am more and more bent (although it isn't due to fatigue or unbalance; it isn't due to any material reason): I have the feeling that the present part of the body (the part that belongs to the past, rather) is growing smaller, while, on the contrary, I—my consciousness— I am so vast, so tall, and so powerful. . . . I don't know how to explain; it's a funny sensation. It's like lugging an old piece of baggage around. But it isn't unwilling; it's just that the difficulty varies, you know, so it takes more or less time. There is just some lagging.*

Time, always time.

*But I'd like to have the answer,* Mother continued. *The problem is beginning to interest me!* [There was always that touch of genial humor running in the background.] *Is that residue going to. . . ? That is not the right question—it's a question of TIME. Given time (Sri Aurobindo spoke of 300 years), given time, EVERYTHING could change. But there is the wave of habits and the easy solution of taking this old clothing and getting rid of it: "Get lost. I don't want you anymore." It's disgusting. Because the process isn't fast enough, you turn to it and say: "Get lost! Go back to the earth." It's disgusting. And I can feel it in the atmosphere, in the collective thought. People write to me, "I hope you live for a long time!" and all the usual stupidities. It makes for a . . . difficult atmosphere. I look at this body, and sometimes* it says (when there's too much incomprehension around, when those around are too utterly uncomprehending), it says, "Please let me go" ("it," that is, the part that's still unconscious, too unconscious and not receptive enough); it says, "All right, let me go. Never mind, just let me go." But not tired or disgusted. So I say to it: "Absolutely not!" [Mother took the tone of voice one uses to talk to a child]. It's a question of patience, you know. . . .

For whom? For her or the others?

But there was no question of letting the old residue fall to dust.

*A question of patience. . . . What is going to happen? . . . I don't know. We'll see. In any case, you, for one, will know. You will be able to tell them: it isn't as you think it is! . . .* [And Mother laughed ironically, as if She could see the whole "atmosphere."] *I would tell them myself, but they won't hear me! I don't know. I don't know what's going to happen. What is going to happen? Do YOU know?* And I looked at her, and it was so obvious! "One day, it will be glorious." *When you do something for the first time, nobody can explain it to you.*

She was so alone, so alone.

At the frontier between decay and something else.

But one morning in February 1970, Mother suddenly remarked, *There's a curious thing: I don't sleep and I am not awake. It's neither one nor the other. I am in a sort of new state, whether I am in bed or sitting in my armchair, it doesn't make any difference. . . . It's something different. And I don't sleep! What is it? . . . I don't know. There's something. . . . Is that possible? . . . And I am not out of my body. Or else this body is replaced by another one? I don't know. But everything is different.*

Is that possible?

A new state in matter. . . .

Mother's last mystery.

All the data were clear. There only remained the unknown.

Mother is ninety-two. She had three years left.

FIFTEEN

# The Supreme Door

There had been so many experiences since the time a little girl sailed over the stones at Fontainebleau, spoke to the big python, listened to the tale of the mummy in the Guimet Museum, and beheld, in the transparent pages of her textbook, a living history that seemed so very familiar to her: the strange kinds of matter in Tlemcen, the precipitous falls, the trips outside the body, the first encounter with death, and the impressionist explosion of light, the revolution of the atoms, the revolutions flaring up from the Yangtze Valley to Moscow, then a postern adorned with a vine of "Faithfulness"—a long, immense faithfulness to the earth's history, to the old march of matter toward its fulfillment. Throughout all ages, all sufferings, black or golden experiences, in Thebes's hallways or the dungeons of a Palazzo Ducale, that same quest for a truer, wider, freer world, without borders, without religion, without police, for a "sunlit path" for the earth: *I say there's no need to suffer.* And at the top of a little stairway leading to a large, white veranda stood He who said, "The world is preparing for a new evolution." She has traveled far, this Mother, traveled to the last frontiers in this prison of matter She so much wanted to open up for humanity, in that dungeon covered with a golden carpet where all the suffering of a world clinging to its falsehood and opacity kept assaulting her. And She laughed. She opened wide her motionless eyes upon an already changed earth. She said strange things while She gasped for air, and desperately pulled on the line to make herself heard on this side. Imaginations? *I prefer this imagination to yours,* She said simply to the evangelists of death, those who believe in the tomb, in science, in the prison forever. Oh, how much She wanted to bestow on men, give back to the earth its own creative power, its "imagination of truth," wrest from matter its own miracle. The miracle of Truth, for there is no other. And love forever, in everything, through everything—the good, the evil, the small, the large, the white and blue and black, and all the samples of the world's suffering.

*At least, we'll have tried.*

## The Old Way Is Dying

And I say She did not fail, any more than evolution can fail. It is just that there is something we do not understand—not yet—an unknown piece of data. Or something that is still not manifested, but ready, like the chick inside its shell. Or perhaps like the caterpillar inside its cocoon. If a person has never seen a chick emerge from its shell, how can he ever imagine that that calcareous crust holds a bird? It took us a long time just to evolve from Neanderthal to Lascaux, and from a barren cliff to the first little lavender plant. The secret always lies in discovering what is *there,* inside the evolutionary shell. That unknown piece of data is *there,* in Mother's last years. This is what has to be deciphered now. It is not really Mother's secret; it is *our* secret as a species evolving toward—just what Mother sought.

"A strange state," as She said. All we can do is observe the unfolding of the phenomenon in a "clinical" way and see if, by chance, we stumble upon a whisper of the future. According to medical standards, She was blind and deaf—a walled-in world, the absolute material shell, a sort of living tomb for any of us—and yet, She heard my least word and saw better than I the slightest vibration of my own being or objects that informed her where they were; She could hardly walk, and yet She went everywhere and knew all the travails of bodies and the movements of the world and circumstances; She no longer slept like us, barely ate, and yet there was that extraordinary energy around her; She seemed to be sinking into senile torpor, and there was that crystal-clear lucidity which saw everything, understood everything, smiled, eyes closed, at all our stupidities or at a pure little flame; She forgot everything but made every gesture as if infallibly, precisely knew each person's duplicity or truth, each thing at each minute. She was a paradox of consciousness and vision and wideness within walled-in, annulled matter, like a shell growing increasingly thinner. A triumph of consciousness over matter, or perhaps the extreme product of the long evolutionary march. It seemed as if all there was to do was spread the wings of an immortal consciousness grown to the size of the earth and beyond. Who would care about a digestive tract at that stage? Who would bear all the countless troubles of a body that could not even recline in bed because it was so bent, when She could have cut the current in a breath? She knew perfectly well how "to die" at will. "Anybody else would have left a hundred times," I told her one day, "rather than stay here and endure what you are enduring." It was completely unhuman, no doubt about it. All that for a little shell. The "shell" was our whole terrestrial envelope, the evolutionary stake. A triumph of consciousness *over* matter or a triumph of consciousness *within* matter?

She had reached the ultimate mineral, or atomic, layer, that first beginning of the shell and of all shells. She was face to face with the beginning of the world in her body, on the scene of the original hardening of matter—just a thin little peel bathed and steeped in a dazzling but apparently motionless flood of energy. A "permeation" that looked and felt just like a mighty churning, a pummeling, a devastation in the flesh, something perhaps not unlike what takes place inside our cyclotrons, only bearable because of that kind of "dilation" of the body consciousness, which caused it to widen to the size of the universe, as it were. One second of slackening, to slip back for one second into the "hole," as She said, into that "my body," that perception of "I" move, "I" eat, "I" speak was sheer hell. It meant being back into the living devastation. There was no longer any layer of unconsciousness and darkness to protect her against "that assault of ether and of fire,"[1] as Sri Aurobindo called it; there was only that body, pure, without walls, unless it was itself perhaps the last wall undergoing its own transmutation. And on the other side, a universal, *physical* consciousness, another state, another, unspeakable way of being, no longer governed by the laws of the tomb, no longer in need of eyes, of ears, of memory, or even of a body to move with, a way for which the very pain of the body was a sort of unreality—yes, perhaps a different kind of time, a "moving eternity" in the depths or behind that atomic film, a different state *of matter* behind that hardened crust. On one side, the hole of death and pain; on the other, universal life, free of any possibility of pain. A "funny," two-tiered *physical* life, a paradoxical duality. *The life of this body is a miracle,* She said in April 1970, after a series of heart attacks. *I mean, if it weren't what it is and how it is, anybody else would be dead by now. But if you knew how strange it's getting! The body is conscious and says, "Actually, it would make a difference mainly to others* [if Mother 'died']!" *To me, it's ... But they, you see, still live in that sort of illusion that one is dead because the body has gone; and even this body doesn't know which is true anymore! For it, matter should be the truth, but even IT isn't too sure that's the case! There's the other way of being. So the body is starting to wonder. . . . It knows the old way is outmoded, but it is starting to wonder how things will be. Sometimes, strangely, that comes—it comes like a breath, and then it disappears again. Like a breath of a different way of looking at things, a different way of feeling, a different way of hearing. It's like something that comes very near, only to be veiled again.* Actually, this new functioning had manifested itself for quite some time already, but each time it was a discovery for Mother. . . . *And the body feels pain, a funny kind of pain, a very funny kind of pain: my body moans, literally moans as if it were in terrible pain, and all the while it says to itself, "Ah, this is bliss!" And it moans! The two together, you understand? And it depends on a tiny . . . something like an act of will, but it isn't will. I really don't know. It's something new.* The "tiny something" was the imperceptible shift from one position to another, from one kind of time to another. . . . *The body moans and feels it's in pain, then a little something happens, and there's no longer a feeling of pain, but it's not at all what we call "bliss"—we don't know what it is. It's something else. But it's something extraordinary. New. Absolutely new. . . .* It is really the beginning of a new state in matter. *It's no longer—it is visibly NO LONGER the same body consciousness it used to be. It's no longer what it was: the relations are no longer the same, the way of hearing, the way of speaking are no longer the same. . . . And that's not it yet. . . . Oh, it's on its way to something, but it's not there yet. But the presence of the Grace is something absolutely marvelous, because, as I see, as the experience unfolds, if I were not given the real meaning of what is happening when it happens, it would be nonstop agony—the old way is dying. Of course, there has been a whole yogic preparation, but for the body . . . it's truly a constant miracle, you know. It couldn't be withstood more than a few minutes. And yet it goes on and on and on. . . .*

It would go on for three years.

Three years of constant agony during which slowly, imperceptibly, but irresistibly there was born a new state of matter, which was no longer life as we know it, not death as we conceive it; no longer time as we measure it, not the eternity where one falls asleep; no longer the pain that causes one to faint, not the yogic bliss in which one faints in another way; no longer matter that dies of a heart attack, not the non-matter in which one soars transcendentally.... Something else. A completely new state.

And in the midst of it all, a shell between two worlds or inside two worlds. And Mother added this, which is definitely very mysterious: *We think that this [the body], the appearance, is the most important. For the ordinary consciousness, it appears to be the most important. It will be obviously the last thing to change. And, for the ordinary consciousness, it seems it will be the last thing to change BECAUSE it is the most important—it will be the surest sign. But that's not it at all! That's not it at all. The change IN THE CONSCIOUSNESS, which has taken place, is what is the most important* [the change of position with respect to time]. *All the rest is just the consequences of that. For us, it's only when this [the body] is visibly capable of being something different that we will say, "Ah, now the thing is done." But that's not true—the thing IS done. This [the body] is just a secondary consequence.*

And that raises a host of questions. What is so decisive in this new state or new position as to make the conditions of the shell so negligible, or at least secondary? Is the shell's transformation—which means the transformation of all our visible terrestrial matter—superseded by another fact? Is the "transformation" something else altogether?

"If we knew what it is," Mother had said ten years earlier, "it would be done already!" Perhaps She was nearing the point, or the position, where She was beginning to know what it was?

As for me, I do not have the faintest idea.

Perhaps She will give us a clue to this last mystery?

## The Mystery of the Contradiction

The clinical picture continues.

A growing, increasingly acute paradox. *It's hard. English-speaking people would say, "It's not a joke." Absolutely everything is getting disorganized. It is clearly getting disorganized TOWARD a higher organization, but . . . But nothing works the way it used to. The body can no longer eat, no longer . . . It's a very strange sensation: there's no longer any of the relations there used to be, none whatsoever, whether it concerns the body with itself or the body with others or anything—it's all completely gone. And once in a while, a fleeting little breath, you know, a little something that's—I don't know how to describe it—delightful. It isn't pleasure, it isn't joy; it's . . . a passing little breeze that's quite special—delightful. The next minute, it's gone. All of a sudden, the body feels a kind of peaceful and luminous and . . . truly lovely rest—the next minute it is in pain all over again. And everything is like that. There's a sort of identification with everything, which is far from being pleasant, though it isn't really unpleasant, but . . . it makes for a strange kind of life. One moment, the impression that you depend on nothing, that you are an expression—how shall I say it?* [Mother smiled]—*an expression of the Lord, and you depend on nothing; the next minute, the impression that you are nothing but a semiconscious movement in the midst of general semiconsciousness—quite unpleasant. That's how it is, and it's constantly that way. At times, things are so . . . almost repulsive that you feel like screaming—and in fact, if you are not careful, you would scream; at other times, everything is so peaceful that you feel you are entering an eternity. So, as you see, the only thing one can do in the midst of all that is be calm!* Fortunately, as Mother said, there was no propensity toward unbalance. Fortunate chromosomes from Mathilde. *And along with all this comes the awareness of everything that people think—of everything they think of IT—of all the ... Oh, it's so deplorable!*

Mother said that the change was DONE, but clearly something was not yet done (?). Or was it? And the contradiction kept growing, so to speak; the swings from one state to the other grew ever shorter, like an electric arc ceaselessly illuminating a hole of darkness and traveling in one direction and then in the other. And the experience always seemed to thrust her in the same direction, to compel her to find the solution. *There's this experience of a tiny little shift, a tiny change of attitude— which isn't even expressible in words—and in one case, you are filled with divine bliss, while in the other, in exactly the same circumstances, everything becomes almost a torture! And this happens constantly. There are times when the body would howl in pain, you know, and ... with a tiny little change, almost inexpressible, it becomes blissful. It becomes ... something else— it becomes that extraordinary presence of the Divine in everything. And the body keeps going from one to the other, like a sort of exercise, of struggle of consciousness between the two. And all the painful vibrations are as if sustained by the mass of the general human consciousness—that's the problem!* Something that *sustains* pain in the world. Which almost loves it. One day, as She was suffering from a tooth infection (once more), which She had managed to "densify," stop, reduce the swelling, *someone approaches the body, thinking, "Oh, poor Mother! How She must suffer"—it returns immediately!* ... Indeed, She *must* suffer. The cellular contagion was right there, constantly there. And at the same time, progressively, so was the direction of the solution *within* the contradiction: *And this body passes from one to the other,* She continued, *and sometimes almost both states together. The result, for the appreciation of ordinary things (I mean, of life as it is), is a perception of general mad-*

*ness, with hardly any difference between what people call mad and what they call reasonable (the difference they draw between the two is truly comical!). There is a whole world of simultaneous perceptions which is truly impossible to describe. Something that experiences innumerable things at the same time, while the capacity of expression has remained what it was, namely, incapable. But that "passage" is the most constant work: there are no more ideas, no more feelings, almost no more sensations, just that sort of shift—a shift that results in SUCH a difference, and in complete immobility!*

This was the increasingly pronounced phenomenon: a sort of cellular eternity or cellular immobility in the very midst of the contradiction, which perhaps even resulted from that contradiction... to the point that one was left wondering whether it was not in fact a mistake to think of a side of death and suffering, and a "true side" of eternal bliss—maybe there was another location, made up of both sides, as it were, a third... incomprehensible state? And *in the body,* mind you, in matter. It had nothing to do with ethereal states; it had to do with a raging toothache or a body in the process, or so it seemed, of dying. A state that completely defies what we call the natural state, of the world of Nature, and yet which was part of the physical world. Maybe a new physical nature? The physical nature that controls a fish and that which controls man are clearly quite different, and yet they are the same. *All the functions that used to take place naturally, in accord precisely with the forces of nature, are suddenly gone, finished! They are withdrawing. And then comes... something... which I call the Divine (perhaps Sri Aurobindo called it the Supramental, I don't know, it's something like that), which is tomorrow's realization (I don't know what to call it). When everything is thoroughly disorganized and going really wrong, That consents to intervene. Not a very pleasant transition at all, you could say. Acute pain, impossible to eat, impossible to... etc. Obviously someone had to do it!* She was making the road. What road? To where? In what? Nobody knows. Can the caterpillar describe the road to the butterfly? And at least the caterpillar is comfortably nestled in its cocoon—but how about doing it with eyes wide open, without hibernation? Sometimes, I tried to ask her questions (less and less because my heart ached), as I so much wanted to make sense of it all: *I don't know!* She exclaimed. *My body is undergoing the process.* And that was exactly it. *It feels a work of transformation is taking place. There are moments when it feels it's impossible—it's impossible, one simply cannot exist like that—and then, just at the last minute, something comes, and there's . . . there's a Harmony absolutely unknown to this physical world. A Harmony—the physical world seems dreadful in comparison. But it doesn't stay. But the perceptions are clearer and clearer, more and more luminous—more and more encompassing. It's truly like a new world seeking to manifest.*

And we have to admit that we do not understand—the end result, yes, we can understand; the other side, yes, we can imagine. But the transition, what *creates* that new state? The mechanism? We can only note that it is only when everything is completely disorganized that "That consents to intervene." Of course, the caterpillar must be completely disorganized for something of the butterfly to consent to intervene. The caterpillar must reach a sort of state of total contradiction. And what happened *in* that contradiction? Always the same movement of densification or "eternization" (maybe that was the particular form of hibernation for a human transformation, with eyes wide open). But was it only that—a means of effecting the transition—or the new state itself? We are obsessed by the idea of this old body, which, we assume, must be "transformed"—it will certainly have to be transformed; it is not meant to remain this old, stiff, rigid garment—but should not its very "breathing" conditions change first? A physical way of breathing that causes a certain death or decrepitude versus another way that makes for life without death and decrepitude. If that particular way, or particular breathing environment, is established in physical Nature (I mean, of course, a different breathing process having nothing to do with oxygen or nitrogen), then the problem of transformation actually disappears—it is a "secondary consequence," as Mother says—it will naturally and gradually take place in the species at large. And we recall the story of the axolotl, that sort of half-worm, half-fish animal, blind and colorless, living in the mountain lakes of Mexico and reproducing, dying, reproducing in a larval state for centuries perhaps, until some scientists took a few specimens back to their laboratory to study. To everyone's surprise, within a few days, because the conditions were different, those worms took on colors and changed into adult salamanders (which they named ambystoma), which proceeded to reproduce normally as salamanders! The scientists were confronted with the extraordinary fact that the axolotl was a larva living, reproducing and dying *without leaving the larval state*—but when the environment is changed, the inexorable groove is broken and the larva becomes a different being living and reproducing in its own way. A change of environment. A change of air. And everything changes. A new breathing environment (it is hardly a metaphor) is what needs to be created. Maybe it was this new environment that was trying to emerge or see the light of day in matter through the bit of matter She represented? It is like a new type of air, breathable or absorbable by all those who have reached the required level of aspiration—one must of course aspire in order to respire! A certain quality of aspiration in humans should open to them the door or the window to that new *physical* air. Naturally, it is not the physical of nitrogen, but it is physical nonetheless compared to the subterranean air we breathe, or compared to the air of erect goldfish we soak up inside our rather muddy bowl.

Was this what was happening in Mother's body?

Of whom are we the larva?

Was that acute, painful, hellish contradiction not the very instrument of the new mode of being or breathing? Indeed, one day in 1970 She had made the following remark: *It's become quite interesting. I spent the entire night with Sri Aurobindo, but with a world of explanations. He made me understand a multitude of things, but quite—well, just extraordinary. A detailed demonstration of the difference between the two consciousnesses. Among other things, He explained to me, in a very concrete and practical way, that all illnesses, disorders, conflicts in this material world are caused by the separation of two simultaneous movements—one is the movement of duration (what could be called Stability) and the other the movement of transformation. These two movements are one in the original Consciousness—they do not contradict each other—and I was shown how they're separate here, and that's what causes death. Just because they can't harmonize with each other—they don't know HOW to harmonize (they can, but they don't know how to). One is the movement of transformation, and the other the movement of stability. . . . In other words, they are like Eternity and the Becoming, the movement of progress and immobility. When they are not in harmony where needed, that produces a loss of equilibrium, and the being dies—things die, everything dies because of that. And it's so simple, so obvious once you experience it! . . . One could say (almost say) that if the two can find their SIMULTANEOUS equilibrium of existence, that recreates the Divine. He is in us, but not harmonized.*

And we are again confronted by that supramental, cellular time which combines perfect immobility and dizzying movement. Indeed, this was the bodily, physical state—the "environment"—trying to become a permanent reality in Mother's body through and because of the acute contradictions that threatened to swallow her into death at each instant. This was the other breathing.

And one day, the mystery of the contradiction (which I took for the transition to the "other state," the other side, the "blissful" side) appeared, or seemed to appear (?), as the very condition, or very site, of the new state—there was no other side to go to! For I used to complain to Mother and, referring to my own experience, which seemed psychologically to follow what She experienced physically, I said to her, "The farther I go, the more I uncover contradictions in myself—acute contradictions. They feel as if they were impossibilities." And She answered: *No, not impossibilities. It must be that you have to go deeper or higher, to the place where they become one. That's how it is: oppositions become increasingly intense until we find the place where they . . . where oneness can be established. One has to go deeper and deeper or higher and higher (it's the same thing). All our old ways of understanding things are worth nothing— NOTHING. All our values are worth NOTHING. We are on the threshold of something that's really marvelous, but . . . we don't know how to keep it. I have never, ever had such a strong impression of knowing nothing, of being capable of nothing, of . . . being a pile of dreadful contradictions, and yet I know, I KNOW (without words, deep down) that it's only because I am unable to find the place where . . . things are harmonized and unified. And I have a strange feeling of torture and bliss almost at the same time—almost at the same time. There you have it.*

Clearly a new state which is not bliss, not torture, not good, not evil, not life, not death—something else . . . which combines all that together to make a new substance, a new being, a new air. All our values—medical, spiritual, moral, legal and scientific—are worth nothing, because each is its own marvelous thing inflated and magnified. Even our ideas about transformation may be human fantasies. . . . There is another air. Another environment. There is a place where everything is one. We carry within us all the contradictions needed to reach the place. Mother too.

Will She reach the place in the body where the contradictions of the earth are reconciled—where life and death are reconciled? Or, rather, where they combine and dissolve into something else which is the third state we are seeking? A state in which transformation is no longer a sort of personal feat of strength but the natural consequence of a certain way of being and breathing. Like the axolotl in its new environment. Forty years earlier, in 1930, Mother had said, The true change of consciousness is one that will change the physical conditions of the world.[2]

## The Layer of Carbon

Then the great blows fell.
Truly a furious demolition.
As if the passage were carved out in her own body.
The first one came in August 1970, when her personal attendant, who had served her so many years with great fidelity, had to stop her work. Already, a year earlier, She had lost her smiling treasurer, Amrita, then Pavitra, her general secretary, one of the only pure and reliable people around her. She was increasingly alone before the pack. For a whole month, during those days of August 1970, her body fought with death. It was a repetition of the turning points of 1962 and 1968: *The body is left on its own. . . . It is always the body that must find the solution. And each time, it is emptied of whatever it may have acquired in the meantime. It has to reach the place where the thing is done.* I still hear her breathless voice (her lungs were affected; She coughed continuously): *If I can remember when it's over, I will really have something interesting to tell. But I am not sure I will remember. . . .* And a few days later: *This little body is like a point, but it has the feeling of being the expression of a formidable power. And it is helpless: no capacity, no expres-*

*sion, nothing, and rather miserable. And yet, there's this incredible concentration of power! Sometimes it even has trouble withstanding it, you know. All the experiences have sort of increased a hundredfold. . . . But my legs hurt. It's 24 hours a day, you see, and no possibility to rest. That's the trouble. . . . If I let myself go, I would scream. . . . Terrible. . . . That night, I said to myself, "This is what hell is like." Terrible—it's terrible. I don't know why I had to go through this. . . . Because it meant that death wasn't a solution, you see! And that's frightening.* She was reaching the point, the point where death was not a solution. Because there was no more death as such in her body, no more oblivion, no more "I close my eyes and I get out"—get out where? She was fully conscious. Each cell of her body was fully conscious. There was no more "other side" of oblivion, nowhere to escape from that. Can anyone understand? . . . *It's so horrible, I am tempted to say: Pray for me. Lord . . .*

But a disciple who had a remarkable gift of vision had, at the same moment, seen the following: "Mother was going down, down, sinking into the earth, and She was as if completely covered by a layer of carbon. There was light wherever She was, but the thread connecting her to her origin was extremely tenuous—a very slender thread going through that carbon layer. And sometimes the contact was lost—the thread disappeared—and Mother was in difficulty." She was reaching the root of death. Total asphyxiation. And *at the same time,* that formidable Power one felt around her, almost crushing, growing ever more formidable as her body was annulled. *I had the feeling I was all the suffering of the world . . . felt together. I don't know how to put it. That feeling of being crushed is not yet gone. It's like something that keeps you from breathing freely. And then I became aware (and that was really frightening), I became aware of all that Sri Aurobindo had to suffer physically. That was one of the most . . . difficult things to bear [Mother's eyes were full of tears] . . . our physical unconsciousness next to that, and the sort of physical torture He had to go through. That was one of the most difficult things.* Sri Aurobindo had reached the same point while He sat there, in the middle of all the little recalcitrant axolotls. Now She understood. And they "treated" him ... according to the medical manuals. "But isn't the purpose of that suffering of the earth to call the supreme Consciousness down to that level, all the way down at the bottom?" I asked Mother. *Yes, of course. That's what I tell myself, what I am trying to find. There's something to be found....* And She stopped, out of breath, as if suffocating: *That's it, there's a place, a sort of place, where there is such a dreadful anguish.... And that's constant. It's here [Mother drew a line over her chest]. It's right here. And then there's a sort of prohibition to exteriorize myself....* She could no longer leave her body. *As if I absolutely had to find something.* "We will overcome, Mother." *Yes. I am absolutely convinced it will be overcome, you see, but... has the time come, that's the question? And that, that doubt, is what tortures me.*

Then She recovered. She started to walk back and forth in her room again, indomitably. The cough stopped. The invasion of people started up again.... *If only I could have purged the world by having those days of horror, then it doesn't matter, I don't mind.*

But She had not yet found the "something."

Five months later, just enough time to breathe a little, the second blow fell, even more radical than the first, if that is possible. That place of the change had to be reached. This time, it was a paralysis of the left leg (due, it seems, to a blood clot). I saw her a month and a half later: *What ties me down is this paralyzed leg: the lower part, from the knee to the heel. So naturally you become stupid; you can't do a thing! ... It's coming back slowly. There was a time when it was total; it was as cold as ice; there was no circulation. But it was far from being an innocent paralysis! For at least three weeks, there was constant pain, night and day, 24 hours a day, without any fluctuation whatsoever. It was as if everything were torn out of me. . . . This one [the right leg] was also nearly caught, but the day it happened, I concentrated tremendously, and I walked for a long, long time to keep it from being caught also. . . .* She would walk right to the end. *You could say I was nothing but a scream. It lasted a long time. It lasted several weeks. I didn't count. Then, gradually, came periods of quiet when the leg was not in pain. And only in the last two or three days has it seemed to be back to normal. You see, it was so—it was the whole problem of the world!—a world that was nothing but suffering and pain, and then a big question mark: why? . . . I tried all the remedies I knew: changing pain into joy, suppressing the capacity to feel, turning my attention to something else. . . . I tried every "trick" in the book—not a single one worked! There's something in this physical world as it is that isn't . . . that isn't yet open to the divine Vibration. And that something is what creates all, all, all the trouble. The divine Consciousness is not perceived. There are masses of IMAGINARY things (though very real for the sensation) that exist, while That, the only true thing, is not perceived.* And here we come to a very critical point, as if all that pain of the world, so very concrete (paralysis is not something "innocent" or imaginary), were only the result of a false consciousness of matter or in matter. It is a critical point because a consciousness *can* be changed. One might be inclined to think that death cannot be changed, nor can cancer or paralysis, but consciousness *can* be changed.

There is a consciousness to change in matter.

There is a mode of vibration to change in matter.

There is a true vibration that changes everything—including death.

Would a conscious axolotl, with the gift of vision, not say that its subterranean environment is a false environment, an unreal reality? An opaque illusion? Only, in the present case, the true environment *already* exists. We are in it. It is *right here*.

But one year earlier, in 1969, Mother had had an exactly similar experience, which I had not quite understood then, with the same conditions of physical disintegration and pain as in 1970. All of a sudden, in the middle of the disintegration and suffering came a light, "something," which had *physically* changed everything. As if all that pain was not real. A kind of *bodily* duplicate of the Buddhistic experience of Illusion—but instead of being destroyed in the consciousness above, while the body continues to suffer below, it is destroyed *in the body*. It is in the body that a certain illusion must be destroyed. *Never, ever, in its entire existence, has the body felt such a total and profound pain as that day,* said Mother. *Oh, it was something . . . You know, separation, and nastiness, cruelty, suffering, and then disease, decay, death—destruction (it's all one and the same thing). Well, the experience I had was the UNREALITY of those things . . . as if we were stuck in some unreal falsehood, and everything vanishes when we get out of it—it doesn't exist, it no longer is. That is what is so frightening! The notion that what is so real, so concrete, so frightening to us is all nonexistent. It's just—we are stuck in Falsehood. Why? How? What? . . . And at the end of it all, Bliss. Poof, everything vanished! As if all that, all those dreadful things, had no existence.* Something in the body that causes evil, death, destruction to become unreal. *And all the means that we could call artificial—including Nirvana—all the means of getting out are worth nothing,* Mother added. *Beginning with the idiot who kills himself to "put an end" to his life; of all the stupidities, that one is the greatest. From that all the way up to Nirvana (which is supposed to help you "get out"), it's all absolutely worthless. Those things are not on the same level, but they are all worthless. So after that, when you really have a feeling of perpetual hell, all of a sudden . . . it's just a state of consciousness, nothing other than that—all of a sudden, a state of consciousness in which everything is light, splendor, beauty, happiness, goodness. . . . Here I am. See? It shows up, and poof!—it's gone.* And Mother wondered, *Is that it? Is that the lever? . . . I don't know. But SALVATION IS PHYSICAL—not in the least mental, but physical. I want to emphasize that it isn't in escaping—it's HERE. There's only ONE exit from all this, only ONE—ONLY one, not two. There's no choice. There aren't several possibilities; there's only one—it's . . . the supreme Door. The Marvel of Marvels. Everything else—everything else is just impossible. And it isn't veiled or concealed or anything; it's here. Why? What is it, in the whole, that keeps us from experiencing it? I don't know. It's here. It's right HERE. And all the rest, including death and all that, truly becomes a falsehood, that is, something that has no existence.*

Another environment where all this does not exist.

Mother was coming to the central point, that knot of Falsehood and pain, that layer of carbon. The absolute contradiction. The axolotl's suffocation. And right in the middle of it, suddenly, something . . . bliss, the Marvel. The lever. Something changes in the vibration of matter, and everything is changed—including death: the old groove of the axolotl.

*There has been a tremendous change,* noted Mother after the days of horror in 1970 and 1971, *but nothing can be said about it. . . . The whole body is no longer the same. It's really as if it had been prepared for another type of consciousness, because there are certain things—its reactions are entirely different, its attitude is different. I went through a period of total indifference where the world didn't represent—didn't mean anything. Then, from that, something like a new perception gradually emerged. . . . It's just in the process. And I noticed how the so-called catastrophes or calamities or misfortunes or difficulties, how they all come JUST at the right moment to help you, JUST when you needed to be helped. You see, everything in the physical nature that was still part of the old world, of its habits and ways of doing and being, of its ways of operating, could not be handled in any other way than this—illness. I can't say it hasn't been interesting.* Exactly like Mother: She was studying a phenomenon. And after a long silence during which She looked straight ahead, She suddenly added, *The world is in a dreadful state.* It too, inside the carbon layer. "Yet, I have never felt that the moment was so close, so very close, as I do now," I said to Mother. *Yes, yes, very close. . . . I think something will have been achieved from the general standpoint—it wasn't just the difficulty of one body or one person—I think something has been accomplished in terms of preparing matter to receive as it should. . . . It will come. I don't know if it will take months or years for the result to become apparent.*

Gradually the result became apparent.

A new passage was carved out in her body.

Or a new state.

A state of consciousness that changes the physical world and physical nature—that unravels the eternal contradiction of life and death, of pain and bliss, of escaping to heaven or into Nirvana while the earth rots. Truly a new evolution.

Salvation is physical.

If that is found, it is obvious that physical transformation naturally and spontaneously follows, just as the salamander naturally and spontaneously follows a little axolotl awakening from its layer of black mud.

SIXTEEN

# Overlife

This book is a challenge. Mother's entire undertaking is a challenge. Sri Aurobindo's silence was no doubt great wisdom, but I am trying, nevertheless. This is my own challenge, and I am not about to stop poring over those stammerings from another world. I am struggling with Mother's forest as She fought and struggled with the unknown. All the secrets are there, but without a name—what formula can possibly trap the Amazon?—one can walk there, lose oneself there, roam in every direction, and everything is as if laden with meaning. With Mother, one felt that She walked with a perpetual open sesame. Each thing had its open sesame, the slightest trifle—everything was a perpetual open sesame. The great Door can indeed open at any page, at any line of that *Agenda*. There is no need to understand. That may not even be necessary as much as to seize that straight little vibration which pierces every appearance and opens the new world like a sudden cascade of laughter amid hopeless banalities and darkest contradictions.

We do not know how to read what is there, all there.

Mother is truly the One who un-covers.

She used every minute of her life, every circumstance to un-cover. Until the very end. She never put a definitive meaning on things, because the meaning was to keep walking. Nothing was ever attained, fixed—there was always the next step. Strangely, She walked on nothing and made something spring forth at each step. And it was alive, forever new—a sort of perpetual dynamite inside the old crust we lug around.

And now, what was She going to blow up?

No, She will not leave any Gospel behind, no system, nothing we can really walk on—nothing except a sort of decisive hole in the carapace of earthly habits of looking at things and living, and an open sesame that only waits for us to notice it.

All of evolution is meant to lead to matter's ultimate open sesame.

## Two States of Matter

A hole, of course, is something illogical, a break with what was before. And Mother did not have the foggiest idea where She was setting foot—neither did I. So the best course is to continue with the "clinical picture," in the hope that some coherent pattern will emerge. After those days of horror, something had indeed taken birth, but what? "Nothing can be said about it," She said—we can only understand the lotus seed once it has become a flower. The next world will understand Mother quite well. *I am in the kind of state where you know nothing, that's all. And so my only, only refuge is to sort of cuddle up in the Divine, you know, as if. . . To be You, that's all. Do whatever You like with me, that's all. . . . The feeling of standing on the edge of something, and the slightest false move would send you tumbling down. Everything appears different, all the—everything appears different. The nature of the relationship with others is different; the nature of everything is different, but what, what, what? It's the feeling of being on the brink, or on the verge, or . . . in equilibrium—a formidable Power (really formidable; I have examples) and a formidable powerlessness at the same time. The impression, you know, of being suspended between the most marvelous and the most wretched. Exactly like that. I don't know, better not to speak of it. I don't even know where I'm heading—whether I'm heading toward transformation or toward destruction.* She would never know. "I am not told." And what was transformation anyway? There was the wretched thing, and then . . . Nothing in between. Or else something so new it was completely unrecognizable—if one could recognize it, it would no longer be new! It's always the same story: it takes a long time to recognize what is there. *In life's normal circumstances, the body has a sort of stable base such that it doesn't feel uncomfortable, that it can be occupied with something completely different and still remain neutral: we don't notice its existence, and it doesn't need continuous attention in order to be in a . . . reasonably comfortable state. It's an instrument that functions automatically. But in the conditions I am in at present, the minute the body's full attention is not turned to the Divine, it becomes extremely miserable. That's it: the moment it isn't ACTIVELY concentrated, it feels absolutely miserable. And then it becomes terrible.* . . . The "Divine" was concretely the other state, the other type of automatism, the one that made that body walk and rather miraculously last, made it breathe, and if the body forgot that or got out of that state even for a second, there was a sort of instantaneous suffocation—of course, since the old laws no longer worked. To go back to the cage was instantly to face the law of death. And yet, She was right in it . . . without being in it. In the old world as well as in the new. One foot here, one foot there. Torture and bliss. Life, death. *And I don't know if it is particular to this body, but the atmosphere is full of the most absurd suggestions* [catastrophic thoughts everywhere, ceaselessly, right to the end], *and all that only disappears when it is ACTIVELY concentrated.* . . . Truly a foul atmosphere, the constant layer of carbon. Hence: suffocation, easy breathing; life, death, in fact at every breath. Can anybody understand? At each instant, each second of the day, the air of the old species had to be changed into the air of the new—or else die, or get out of it all and go off into bliss for good, far away from this sorry story. . . . *I don't know if it is particular to this body or if it will be the same for all bodies. Naturally, it is very conscious that this is a period of transition, but . . . it's very difficult.* Perhaps She was making another "air" with her little gasps, an air for all those who are also beginning to suffocate. . . . *From time to time, for a few seconds, there comes a . . . perhaps a "sample" of what is to be, of what will be (when, I don't know), but it lasts a few seconds. That is truly marvelous, but . . . It's a very strange impression, like being . . . on the edge—but the edge of what, I don't know.*

And everything was different. Not only did her body no longer breathe in the normal way, no longer move according to normal laws, it was as if suspended between two kinds of air or two modes of being, both physical since it involved a body—a mortal physical and another physical no one knew anything about, except that it was the body's true breathing, its air that does not suffocate, a sort of automatism that impels exactly and functions without friction, "that" which makes it possible "to go on and on"—but her perception of the world was different. One might be tempted to say or think that She lived or saw in another world; but this is precisely what the mystery is: it was not a nonphysical world any more than the laws were nonphysical. It was one and the same physical world, but seen and experienced differently, as if there were another category of physical laws, another category of physical vision, another category of existence within the other one. Perhaps the next world and the next species trying to emerge, still half veiled by the residue of the old species. Yes, between the caterpillar and the butterfly. And sometimes it was the caterpillar, sometimes the butterfly—but the caterpillar and the butterfly are both physical. *It's strange, it seems to be exactly the same and yet it's becoming totally different,* She said. *All of a sudden, in the middle of eating, everything disappears from my consciousness, and quite a while later, I realize that I am there, with my spoon in midair! . . . Not very practical!* "But what happens during the time you go off like that?" I asked. *Oh, it's very interesting! But I don't "go off"; it isn't . . . I am not at all in trance; I am completely awake, in full activity. I see things, do things, hear people . . . the whole time. And I forget—I forget "material" life. And then someone suddenly calls me back. I do not go out of material life, but . . . it appears differently.* And Mother sat looking pensively. *Truly, I think the physical world is changing. It'll be visible probably in a few hundred years, because it takes a long time to become visible to ordinary human consciousnesses. But the feel is there* [Mother felt the air, the atmosphere around her between her fingers], *as if . . . it were made of something different.* Indeed, the world of the butterfly beginning to emerge, or maybe the air that will enable the next species to breathe—which is perhaps already preparing the next species, which we already inhale without knowing it, as it slowly habituates our cells to a new way of being, breathing, seeing. A slow, invisible transformation. Then She added, smiling, *So all the time, something is telling me: Don't say anything, don't say anything! I mustn't breathe a word because people around me would think I am starting to become unhinged.* And I kept insisting, asking, "It isn't just the perception of the physical world that is changing, is it? It's the quality of the substance?" *Yes, yes! It isn't at all the same way of looking at things, not at all. I don't know. . . . But it's funny. And it's PHYSICAL, that's the extraordinary thing! Previously, I used to go into an inner state of being (I know and have experienced them all; I had a conscious life), but all that is over, finished. It's finished. It's as if . . . the physical became double.*

Two worlds one within the other.

Two levels of physical reality.

And one may wonder who or *what* has that vision or perception of the other physical reality, because this could be just another "way of looking at things," as a visionary has another way of looking at things, or even as a being of a different species has another way of looking at things, as the butterfly has a vision different from that of the caterpillar. But what is remarkable in the present case is that the butterfly was inside the caterpillar's body. It was not a different species that had that vision, not Mother having a higher vision; it was the *body* that had the vision, the cells of the body, the very material consciousness that saw matter, and saw it differently, experienced it differently. It did not take place outside matter, with a different type of matter; it took place in matter itself. Indeed, one day Mother had made a very illuminating remark: *For the body consciousness that remains conscious when the body is asleep, the world as it is (externally or superficially) appears dark and muddy—always. In other words, it's always shadowy—you can barely see—and then mud. And this isn't an opinion or even a sensation; it's a material FACT. Therefore, that body consciousness is already conscious of a world ... that would not be governed by the same laws.* The consciousness of matter, of the body, not veiled by the mental crust and the outer, mind-controlled sense organs—the consciousness that remains in the body when everything else is asleep and the outer organs are obliterated, blind, canceled; the consciousness that is almost like that of a person dead to the world—sees and perceives the physical world in a different way, and not only does it see differently, but it follows different physical laws, which we never experience except in a hypnotic state, or in other abnormal states, because they are hidden from us by the entire mental superstructure. This is the cellular level, the one in which Mother lived, from which She saw the material world differently, and which made her body live quite miraculously despite the onslaught of age, heart attacks, and crowds of people.

"As if the physical became double"—the old physical of the mental world, mental vision, mental laws, and the other one. Now, one understands her strange back-and-forth movement between everything that drew her and suffocated her on the surface, which was like death to her, and the other, cellular state. *Every minute of the day, it's: Do you want life, do you want death; do you want life, do you want death? . . . Literally as if it could die at every minute, and at every minute it's miraculously saved. And that's extraordinary. That's extraordinary. Along with a constant perception of world events, as if everything were—as if there were a connection—a connection.*

A universal cellular level.

Two states of matter.

And I wanted to know more about that other state. What was it really like? What happened in it? And She answered me in her slow, hesitating, clear little voice, as if the words had to cross layers upon layers finally to crystallize in little drops, as at the tip of a stalactite: *When I remain like this . . . quiet . . . after a while, a whole world of things starts happening, becomes organized, but it's . . . another kind of reality, a far more . . . concrete reality. How is it more more concrete? I don't know. Matter seems like something . . . uncertain compared to that. Uncertain, opaque, unreceptive. While that something is . . .* And Mother sat smiling. *And the funniest part is that people think I am sleeping! . . . I almost don't belong to the old world anymore; so the old world says: She's done for. I don't care in the least!* And She laughed, then pulled on a little garland of flowers She was wearing around her wrist, which She called "Patience": *Patience. Would you like a Patience?* And She put her garland on my wrist: *I am constantly told: patience, patience. . . . But others, too, must be patient.*

"I almost don't belong to the old world anymore. . . ."

What was going to happen between those two states?

A great illusion of matter, and . . . an unknown reality.

## Tomorrow's Unknown

It is hard to describe the hell She went through these last three years, to say where it was leading and what it was hiding or preparing. There is no clinical picture of pure pain. There was a mystery one sensed increasing, growing more and more acute, almost palpable, something filled with an unknown meaning; it was right there, one could touch it. But what was it? And I hardly dared ask questions anymore. She herself was a mute, ardent, motionless question, interspersed with little cries of pain. And sometimes She laughed, laughed, made fun of that whole hellish contradiction in her, around her, as though only humor, or Love, were capable of bearing it all: *A silence ... that adores.* And the world's suffering, the world's chaos seemed to increase, the world's contradiction, as in her body—one and the same body in transition ... to what? One day, as I felt strangled by that suffocation of the world, as I looked at Mother's body as if one could really see, touch, feel there the whole mystery of the earth locked up inside its layer of carbon, I said to her, "The only solution is for you to have a glorified body, visible to all, so everybody can come and see for themselves: Come and see what the divine looks like!" And She laughed and laughed. *It would be very convenient! Yes, of course! ... Will it happen that way? ... Well, of course, I am all for it, and I would be quite happy for it to happen to anybody; I don't have the slightest desire that it be me!* "Because," I added, "the world has reached such an acute state of suffering and pain that ... it seems the MOMENT has come." She remained silent, looking. *There's a refusal to answer.* She could not say anything. She herself *was* the question, the living question of the world. "Because really," I insisted, 'it's about time for one body to undergo a change great enough to give concrete hope to humanity. Once that Power has sufficiently permeated your substance, you could conceivably pass it on to other bodies that are ready, couldn't you?" *Oh, but that possibility ALREADY exists. I have constant, extraordinary proofs of it. There are small miracles all the time, you know, all the time.*

The small miracles of a new air . . . which looked like nothing.

Lots of small, surreptitious miracles slipping through the web, or the layer of carbon, as if nothing were happening. Maybe there *is* "something" happening, of which we are not aware because we do not know where to look. And the miraculous transformation of an "exemplary" body may then recede before something far more serious, profound and ineradicable, something mutely and painfully being built for the world behind the little cries of that body trapped in the world's black misery.

That is the mystery to be solved.

Mother's "unknown."

There, only our heart can help us grope our way through the ultimate stronghold of Mother's forest. . . . There was the "other state," which made her move, breathe, see, hear—a miracle of every minute—and this one, ours, mortal, suffocating; and between the two, the no-man's-land She furrowed ceaselessly in a sort of frightening back-and-forth movement, with nothing to hold on to. "No one could stand it more than a few seconds," and it lasts and lasts. . . . There are many little seconds in three times 365 days. Those furrows were perhaps the roadway of the "little miracles" to our terrestrial crust? A road was being built, but which one? She did not know. She was building the road. She *was* the road. She was her burning question in between. And sometimes the mystery was overpowering: *It's as if every way of looking at the world presented itself one after another, the most detestable as well as the most marvelous, one after another, like this . . . [Mother turned her hand like a kaleidoscope], and each one comes as if to say, "See, you can look at it this way or you can look at it that way. . . ."* And *the Truth—what is true? It's all that, plus "something" we don't know. For one thing, I am convinced that that necessity to "look" at things in a certain way, to think them out, is purely human, and it is a transitional method. This transitional period seems long, very long to us, but it is in fact rather short. Even our consciousness is an adaptation of the Consciousness—THE Consciousness, the true Consciousness is something else.* . . . But what? *And so, for my body, the conclusion is to . . . cuddle up in the Divine. Not try to understand, not try to know—try to BE. And thus [in that "being"], the power is overwhelming, in the sense, for instance, that one person's illness vanishes (without my doing anything outwardly, without my even speaking to that person, nothing, absolutely nothing—cured), while another comes to the end and topples to the other side. And that other side has become both extremely familiar and . . . completely unknown.* And this is where we are really confronted with a profound enigma. It seems She knew that "other side," the so-called side of death, very well. She regularly went to the place where the dead and the living are together. "Crowds of them," She used to say. At the cellular level, in the body consciousness, there is no barrier. The "other side" is on the same side; it is only the other side of our mental web, of our mental substance. And suddenly, that so familiar "other side" was becoming a sort of unknown. What does that mean? What was happening? What change was taking place in Mother's structure that caused this difference? A new state of Mother? . . . What state?

And Mother continued, *I remember the time when the recollection of past lives, the recollection of nighttime activities was so concrete; that so-called invisible world was absolutely concrete—now . . . Now everything is like a dream—everything—everything is like a dream masking a Reality . . . an unknown and yet sensible Reality.* The "invisible" was becoming as fictitious as the concrete?! Then where are we? All the past lives, the excursions outside the body, the worlds and realms of consciousness—and even this visible matter. What was happening? *You asked me*, Mother continued, *what happens when I am silent and still like this? . . . It's precisely an attempt at: Truth as it is. And not trying to know it or understand it; all that is totally extraneous—to be, be, be. . . . And then . . . [and Mother had such a sweet smile] . . . then something quite curious happens: at the same time—at the same time—not one within the other or one with the other, but one and the other together: marvelous and frightful. . . . Life as it is, as we feel it in our ordinary consciousness, as it is for men, seems such a . . . frightful thing that one wonders how it's possible to live it for even one minute—and the other, AT THE SAME TIME: a marvel. A marvel of light, consciousness, power—marvelous! Oh, a power, such a power! . . . And all this taking place in the body*, simultaneously. A sort of unknown Reality that seemed to cross that no-man's-land and emerge, to mix with the most extreme suffering, the dreadful misery of our condition inside the carbon layer. *So you can be at the same time in the most painful and incomprehensible and absurd life, and at the same time, at the very same time—inexpressibly marvelous. Of course, I can't speak to anybody anymore. I am only telling this to you, because people would think I'm going crazy.*

A third state . . . as if on the other side and here at the same time. And which altered *both* the other side and this one. A new, unknown—impossible—reality. Almost unlivable. An incredible junction causing an unbearable contradiction of pain and marvel—in the body.

What on earth did it all mean? Where was it going?

It was 1971.

Mother did not know.

*I have the feeling of becoming another person—no, not just that: I am touching ANOTHER world, another way of being . . . which you might call a dangerous way of being. As if . . . [and Mother nodded her head several times] Dangerous, but mar-*

*velous. The feeling that the relationship between what we call "life" and what we call "death" is becoming more and more different—different, completely different.* . . . And Mother kept silent, her blue eyes wide open on the yellow copper-pod tree and Sri Aurobindo's resting place. . . . *You see, IT'S NOT THAT DEATH DISAPPEARS (death as we conceive of it, as we know it in relation to life as we know it). It's not like that, not like that at all. BOTH are in the process of changing . . . into something we still don't know, something that seems both extremely dangerous and absolutely marvelous. Dangerous in the sense that the slightest mistake has terribly serious consequences . . . and marvelous.* The slightest mistake? To lose one's way in the no-man's-land that is in the process of—in the process of what? In the process of being filled with that new, unknown Reality? *Our tendency,* Mother continued, *is to want certain things to be true (what we judge favorable) and other things to disappear [like "death"]. But it doesn't work that way. It doesn't work that way. EVERYTHING is becoming different. Different.* Then Mother closed her eyes, perhaps to listen to the beat of that strange new life which did not belong to any side. *From time to time, for a moment (a brief moment): a marvel. And immediately afterwards, the sense of . . . a dangerous unknown. That's how it is. And I spend all my time that way.*

What if it were the world *as it is*?

The true "as it is" in the process of emerging?

It was 1972. Mother was ninety-four.

*The body has the feeling of being suspended between two states—one that men call life, and the other that men call death. And the body has the feeling of being suspended between the two: neither alive nor . . . [and Mother laughed] nor dead! Like that, neither one nor the other. It is between the two. In no-man's-land. . . . And that's a very strange impression, very strange. You feel (it's not a feeling; it's a perception) that the slightest little disorder would be enough to make you tumble to the other side, and that the "little trifle" that tips the scale is made impossible by something you don't understand. . . . Something that kept you from "dying." Yet it would take hardly anything to . . . You just have to keep very quiet.* And suddenly Mother added, *It's obvious that there is an active Will at work to teach the body how to live in a state that is neither life nor death—something else.*

A third state.

But, nevertheless, sometimes She cried out, so gripping was this unknown mystery—gripping, probably like a new air one is not accustomed to breathing. *Clearly, everything is arranged so you can no longer rely on anything except the Divine,* [of course, for She literally walked on nothing, except that "something], *and I am not told what the "Divine" is! There you have it. Amazing! . . .* At ninety-four, Mother no longer knew what the Divine was! *Everything else is crumbling, just the— the WHAT? The Divine, something—what? . . .* And She kept staring so intently that one felt in one's veins, one's heart, one's body a burning intensity of question—it was burning to live near her. Then She continued, *It's like an attempt to make you feel there's no difference between life and death. That's all. That it's neither death nor life (neither what we call "death" nor what we call "life"); it's . . . something. And that something is divine. Or, rather, our next step to the Divine.*

Then everything clears up, makes sense: It is the next step, the next state of human consciousness, a state that changes both life and death. *It's the true consciousness of immortality.* The one that has the capacity to undo death because death no longer exists; it becomes something else. Something that has dissolved into a third state where both sides are one. No more sides—gone. The no-man's-land is filled, and that is where we are going. This is what was being built in Mother's body— tomorrow's unknown. The mutation of both sides. The constant "small miracle." Then, it is now clear, the body's transformation is a secondary consequence: if we breathe that air, everything changes, everything becomes different. Death is our choice, or our incapacity to follow the movement of progress; and transformation is the natural, inescapable result of that new way of breathing. There is no longer life, no longer death; there is something else . . . and that something is divine.

Life divine on earth.

An immortal consciousness *in the body* refashioning the body in its immortal image. What used to be all the way up there or all the way down in matter, beneath the carbon layer, has come HERE. *All the splendors one can experience by going up, by getting out, by leaving are nothing! They're nothing; they don't have that concrete reality; they seem vague compared to HERE. That is truly why the world has been created. It's in terrestrial matter, on earth, that the Supreme becomes perfect.*

It is not death we must abolish; it is not life we must improve. It is something else altogether that changes those two nightmares into a marvel.

Tomorrow's unknown.

Which is *right here*, in the air, waiting to be breathed.

A mutation of death, really.

In a clear little cell that has traveled the long course of pain from original matter is hidden the ultimate key to the two worlds in one.

Matter performs its own miracle.

The eternal traveler finds again his complete eternity in a body, his totality in a point.

We are there. This is the Hour.

And that Hour we did not know how to define, that state we did not know how to describe, one morning Mother caught hold of it and gave it a name. For once She named something. And She named it just like that, casually, in the middle of other things, as She usually did, without giving undue importance to it, because labels were something She, in fact, lived—She was manufacturing our next label (if we absolutely have to have one). *This is what I've learned: religions failed because they were divided; they wanted you to be religious by excluding all other religions; and every knowledge failed because it was exclusive; and man failed because he was exclusive. And what the new consciousness wants is no more divisions. To be capable of understanding the extreme spiritual, the extreme material, and to find ... to find the point where they join, the point where that becomes a real force. This is also what is being taught to the body in the most radical ways. ...* The impossible contradiction giving birth to, or rather becoming, the real force, the other state. And She added, *The step that humanity must take IMMEDIATELY is to cure exclusivism once and for all. They all say: this, but not that—no: this AND that ... and this other, and that other, and so on, and everything together. To be plastic enough and wide enough to embrace everything. Right now, this is what I am running up against all the time in every domain— in every domain, including the body. The body is used to: this, but not that; this or that. ... No, no, no: this AND that. The great division, you know: life and death. There you are. And everything results from that. Well—words are stupid, but—the overlife is life and death together.*

Overlife.

The state of the superman.

And with her crystalline humor, She added, *Why even call it "overlife"! We always tend to lean on one side: light and darkness ... ("darkness," well ... ). Ah, we're so small! To be sure, one could call it "overdeath"! For that "darkness"—the very one that saints, doctors, police, governments and moralists have denied and tried to eliminate, change, improve—is just what holds the key. In the very darkness of that layer of carbon, in the utter contradiction, lies the Force where the two unite and change into WHAT IS.*

The "Sun in the darkness."

Matter's next state.

Divine and immortal matter.

In 1953, almost 20 years earlier, a little girl had had a strange vision, which she had noted down in English in her exercise book without understanding what it meant. That exercise book happened to fall into my hands, and this is what the January 5, 1953, entry said: "I saw Mother coming back from the balcony. The door was open. Pavitra was there. He asked Mother for a 'message.' Mother handed him a drawing and said, 'This contains Life and Death. You can choose what you wish. The person capable of joining the two doors will be saved.' On the drawing could be seen two houses with lovely green trees. Through the trees I saw two doors. They were separated and closed."

And the two doors are now ONE.

In the body.

This is the supreme Door.

SEVENTEEN

# Uninterrupted Physical Life

But Mother's mystery is not over.
The real mystery is perhaps even only beginning.
We have found a nice little label, "overlife," and everything is exorcised, or so we think—and indeed, that air is there, new, light, for all those who know how to breathe (and even for those who refuse to breathe), but the whole process could take centuries. Ineluctably, obviously, that overlife is the next step of our species, as unavoidably as man follows the little lizard. A change of air, or environment, that will modify all structures. Every evolutionary beginning, we assume, looks like nothing. A tiny golden lichen clinging to a bare rock; a scattering of somewhat crazy men living a bizarre life here and there. But can we afford to wait for centuries? In the days of Charlemagne or Louis le Débonnaire, well, one could conceivably wait for centuries, but suddenly life took on a strange acceleration, which has even nothing to do with our machines—an inner acceleration, as if we were thrust into a funnel that sucks us forward; that kneads and churns, builds and destroys three seconds later, that cuts the ground from under our feet, whether scientific, moral or legal, and we walk on nothing, as it were, forever inventing a nothing that keeps falling to pieces. Clearly, all this leads to something that will not take centuries. We are not

rushing; we are being rushed toward something. One would have to be singularly blind not to see that the supramental, or overlife, or whatever we choose to call it, is at the door. Not even at the door—in the blood, in the veins, in the cells. Along with all sorts of little oddities, which start becoming odder and odder and teeming the minute we start paying attention, looking in the right direction. Everything is teeming with little miracles, little coincidences, little happenings, as if the carbon crust were beginning to let strange signs trickle through all its pores. It even looks like a gigantic convergence of signs. Like little animals in a forest: You remain quiet for a while, and they start stirring everywhere. However, we do not know that particular kind of quiet, which is not even the quiet of meditation but a certain quiet in matter, in the body, in the eyes. We say: "Oh, this is as hard as granite; this is a raging toothache; this is a scorpion, and this is my foot tripping on the sidewalk, and this is . . . as usual"—a million little optico-physiologic labels that block reality, the absolute miracle of each thing, the innumerable message that everything brings us. *They haven't found, they haven't found the true path,* Mother once exclaimed [about young people], *because it isn't a mental path. It has nothing to do with going to inaccessible realms—it's RIGHT HERE. For the moment, however, all the old habits and the general unconsciousness put a sort of lid over it, preventing us from seeing and feeling it. We have to—we have to lift it. And it's everywhere, you know, everywhere, always. It doesn't come and go; it's always there, everywhere. It's only us, our stupidity that keeps us from feeling it. There's no need at all to get out, no need at all.* To be sure, a cow looking at a pasture and a man looking at a pasture do not see the same thing. We have to look differently at the earthly pasture. Lift up the mental lid.

And it will be there, everywhere.

But what is most interesting is that the minute we start looking in that direction, it is as if the phenomenon multiplied, as if it were only waiting for a sign in order to burst out everywhere, in a gigantic complicity of everything. So—so one wonders if those vast, leaden centuries could not suddenly melt in a sort of general little twinkling the moment man begins to look in *that* direction. It will be perhaps the time when we burst into enormous laughter, or when we are seized by a wonderment that will make us definitively leave the human skin. All at once, suddenly, we will be there.

This is perhaps when Mother's last mystery will be unveiled.

For there is still a mystery.

Mother's forest is filled with mysteries.

But one has to look with the eyes of the body.

## Another Physical Air

She was slowly connecting the two doors in her body.

We say "overlife" as if the problem were settled once and for all—a kind of new fixed state, admittedly a little strange but definable and nameable. But for her nothing was settled or exorcised, nothing stopped "at some point"; there was always that same shifting and dangerous unknown unfolding, evolving in God knows what mortal or living, or supermortal or superliving, direction and carrying her into a growing contradiction, which would result in . . . what? What seemed clear is that that state created the conditions necessary for the body's transformation—but when? She could not wait centuries (or if She could, the others could not). And transformation how? . . . She advanced completely in the dark, between torture and bliss, disintegration and a transformation that singularly resembled the disintegration of death. And all the while with a clear, lucid consciousness that perceived everything, saw everything, and observed the phenomenon almost scientifically, in the least detail. So we do not have much choice but to go on with the clinical picture (one should rather say the laboratory notebook) in the hope that some pattern will emerge—we are quite interested in that pattern; it is *our* pattern, tomorrow's mystery. It is the beginning of our own road. Then we will travel it as if it were nothing, with electric poles along it, and we will say, "Look, this is so easy!" We will perhaps even say, "Well, this is evolution after all!" Amen. Unless we have had enough of labels and begin to see the living miracle of everything.

The reality of the contradiction was clear and well laid out, painfully laid out: *Truly, the ordinary state, the old state, is conscious death and suffering. Whereas in the other state, death and suffering seem absolutely . . . unreal. That's all.* And it is not just a psychological "unreality," a sort of fantasy of the "liberated" consciousness on its cloud; no, pain stops physically, heart attacks stop physically, and death *cannot act. If I say nothing and stay like this, immobile, in an attitude of perfect abandon, then everything is fine. The slightest thing interferes with it, and I feel . . . as if I were going to die. It's incredible!* Actually, it is not even "as if"; it's a sort of instant asphyxiation. *I feel I have to lift an enormous weight to make myself heard. I feel I have to speak very, very loud to be heard* [indeed, it sounded as if her voice were passing through layers]. *There's a sort of mass—here, it's as if I were underground and had to shout at the top of my lungs to be heard.* But the underground was precisely our outside world, the old state, whenever She had to talk, explain or solve sordid problems. . . . *And it's a strain, a considerable strain. There is a sort of mass, something with an earth-brown color, weighing on me. As if I were buried underground and had to shout to be heard.* Like the axolotl waking up at the bottom of its hole. Thinking I understood, I said to Mother, "You must be feeling the density of human consciousnesses?" *It's the air. It's in the air.* Naturally, for her the physical air was different. There was another physical air.

Another way of breathing in which death, disease and suffering did not exist, did not have access. They were an impossibility *there*. There was a true matter beneath the layer of carbon. And the phenomenon was becoming more marked: *Things have taken an extreme form. There's a sort of rising of the atmosphere toward an . . . almost inconceivable splendor, and AT THE SAME TIME, the feeling that you can . . . you can die at any moment (not "die," but the body can be dissolved). And so both things combined make for a . . . strange consciousness.* She cut a passage through the layer of carbon. Her very asphyxiation carved, bored the way to the other state through this one. She brought the "thing" forth through each of her little gasping breaths. And She nodded her head: *All the old things seem puerile, childish, unconscious; there [in the other state], it's fantastic and marvelous. That's all. So the body has one prayer, and it's always the same: make me worthy to know you; make me worthy to serve you; make me worthy to be you . . . That's all. I can hardly eat anymore; I am not hungry. I feel an increasing force . . . but of a new nature—in silence and contemplation. Nothing is impossible.*

Nothing is impossible *there*.

Transformation is child's play.

It is the obvious, natural, almost inescapable consequence of that state—there are no "laws" there. There is only *the* Law. But . . . the junction with the old crust had to be made. She spent her time making the junction. *Something has to change in the body's vibration in order for the Consciousness to manifest itself undistorted. The body itself, the physical consciousness, is filled with all those falsehoods and illusions and preconceived ideas, and when that is gone, then the Lord can manifest Himself in it. The distortion is what creates . . . misery, you see, which feels now so dreadful to this body. But when that disappears, changes, it's bliss.*

Something that must change in the body's vibration. As if transformation depended only, simply, on a little "something" in the body consciousness that falsifies, veils, distorts. That little vibration which falsifies,[1] She said as early as 1950. No tissue to transform, no cells to rejuvenate, no bones to make supple or regenerate—just a vibration to change. A vibration that causes old age, rigidity, disease, decay—the whole cursed process. Not a hundred things—one. And perhaps that twisted vibration is even what creates the whole layer of carbon. Change that, and the true Consciousness will begin flowing into the tissues, the bones, the nerves—the earth. Then nothing is impossible. A change of position in the physical consciousness. Something that cancels out the little distorting vibration.

But by what process? What does the axolotl in its hole do to breathe the other air?

## Another Physical Rhythm

But, as usual, the remedy was in the contradiction itself. Mother lived the solution; it came out by itself, more and more. The phenomenon that seemed to accompany and almost characterize the state of overlife was an alteration of time. We saw it appear and develop at the cellular level, as if the pure cells, freed from the web of the physical mind, were endowed with a different time—as they were endowed with universality and with such a swift movement of consciousness that they were as though everywhere at once. This is the time of true matter, we could say. The time of overlife. That phenomenon was now going to accelerate and assume greater proportions, so much so that it seemed to permeate Mother's body, visibly taking over the old crust, filling up the layer of carbon in order to change its substance, its constitution, the vibration that falsifies and veils.

A change of time necessarily means a change in the vibration of the body's consciousness, something that goes faster or slower. And this is where the experience was to take a new direction, or perhaps simply follow its direction to the end. *I don't know how to describe what's happening. . . . In normal circumstances, it can last indefinitely; there's no sense of time or fatigue or duration. When the old consciousness returns, it's almost unbearably painful: I suffocate, or I can't breathe, or it's too cold or too hot, all sorts of things . . . which are exacerbated by a consciousness that should no longer be there. My body, you see, is full of aches and pains, but the minute I enter that state, everything stops—there's no longer any sense of time. Time is interminable in the old consciousness, and it disappears in this one! To use big words, I would say: the old consciousness is death, it's like being on the verge of dying every minute: you are in pain, you—it's the consciousness that leads to death. Whereas the other one is life . . . peaceful life, eternal life.* In the body. In the old crust itself.

And the phenomenon was going to become more and more concrete and visible each day: *If I stay still and I enter that Consciousness, time passes extraordinarily rapidly and with a sense of . . . luminous calm. But if the slightest thing pulls me out of it, it feels as if I were being pulled into a hell. Exactly. The discomfort is so great you feel you won't survive a minute, or several minutes, like that. So . . . so you call the Divine. And you have the feeling of cuddling up in the Divine. Then it's fine.* But what happens in that state? "What do you see there? What do you look at?" was my eternal question. And She smiled: *I feel like saying: nothing! Nothing, I don't see anything. There's no longer "anything that sees." But I AM a quantity, an innumerable quantity of things. I EXPERIENCE an innumerable quantity of things. And it's so very, very much that it's nothing!* And She laughed. *It's a state that is vast, vast . . . peaceful . . . and so powerful, in which things are done. That's how they are being* done. But there are no words or explanations—nothing satisfying for the mind! She added, looking at me teasingly. One day, however, She did try to offer an explanation, a mysterious explanation in which I seemed to grasp, without quite understanding it, a secret, a great secret.

She was looking at all those difficulties of the world, those difficulties in the body, that hell, and She was speaking in her hesitating little voice, as if She had to cross through layers and layers of "earth-brown color": *I am more and more convinced that we have a way of receiving things and reacting to them that CREATES the difficulties. I am more and more convinced of it. If we can be in the true Consciousness all the time, there are no difficulties—and yet things are the same.* There is no illness, no fatigue, no loss of equilibrium—no death—and things are the same. The microbes are the same.... *The day before yesterday I was sick as a dog. And yesterday the circumstances were the same, my body was in the same state, and ... everything was peaceful. If I didn't have so much difficulty in speaking.... That explains everything. That explains everything, absolutely everything. The world remains the same, but it is seen and felt in exactly the opposite way. It's like death, you know. It is a transitional phenomenon, which to us appears to have lasted forever (to us it's forever because our consciousness cuts up everything), but when one has that divine consciousness, oh ... then things become almost instantaneous, you understand? I can't explain. There IS movement, there IS progression, there IS something that for us is translated by time; it exists, it's something ... something in the consciousness. It is difficult to explain. It's like an image and its projection. Something like that. All things ARE, but for us it's as if we saw them projected on a screen: they come one after another. Something like that. Whereas Truth is the Divine as a totality—totality in time and space. And this is a consciousness that the body CAN have, because this body had it, and while it has it, everything is so—it isn't just joy, it isn't pleasure, it isn't happiness, none of that; it's . . . a sort of blissful peace . . . and luminous . . . and creative. It's marvelous. Marvelous. The trouble is, it comes and goes. . . . And when you get out of it, it feels as if you fell into a horrible hole—our ordinary consciousness (I mean ordinary human consciousness) is a horrible hole. But we also know why it has momentarily been that way [in evolution], because it was necessary to go from this state to that—everything that happens is necessary for the full development of the goal of creation. Rhetorically, you could say: the goal of creation is for the creature to become as conscious as the Creator. It's just words, but it's along those lines. The goal of this creation is the Consciousness of the almighty Infinite and Eternal, outside of time, with each individual part having that Consciousness, each individual part embodying that same Consciousness. . . . Words are stupid, but that's how it is. That gives you both the purpose and the goal of creation, and almost its METHOD of development. . . . Yes, it is like something that IS, that is in its entirety, and which is projected on a screen in succession. And yet it exists in its entirety—and it is projected on a screen in succession. I have the feeling . . .* [and this is where Mother touched that Secret] *. . . the feeling that I am on the way to discovering the illusion that must be destroyed in order for physical life to go on uninterrupted—that death is the result of a. . . . a distortion in the consciousness. That's it.*

The screen that traps time, the vibration that distorts.

A "distorting magic," said Sri Aurobindo.

Then She added, *Something IS happening, that's all I can say.*

It was Christmas day, 1971.

Remove the screen, and everything is the same, and there is no more death. There has never been any death *there*. A change of consciousness, a change of time that abolishes death. Death was the transition, the long evolutionary transition inside the cage in order to find, spurred and whipped by pain, the total state in the body.

Remove the screen, and the two sides are ONE.

There is a distortion, a false vibration in the body that *causes* death—disease, disequilibrium, disorder. The abominable chaos of this world. An illusion . . . which is yet real. But an illusion all the same. An evolutionary illusion to help us grow inside the cage . . . and then break it. We break the screen, and *physical* life can go on uninterrupted. There is no obstacle. No obstacle. Death was a necessary illusion to achieve total being, free of death, in a body.

Just a little vibration that falsifies. That veils.

A false time in the body's consciousness.

Transformation means a change in the body's physical consciousness.

Then the world will be exactly the same, yet it will be experienced completely differently—without heaven and without tomb.

As great a difference as between mineral consciousness and animal consciousness.

Something that must change in the physical vibration of the earth.

That is perhaps what is taking place without our knowing it: another rhythm in the body's cells (not in our head; the head cannot grasp it—but the body can). There is a mineral rhythm, a vegetable rhythm, an animal rhythm—and there is the other rhythm.

This is what was trying to materialize in Mother's body, and through her in the body of the earth.

*It's as if another time had come into this time.*

"Something is happening, that's all I can say."

## At Any Second

And we wonder if that mysterious screen is not present everywhere, right before our eyes, so utterly visible and simple that we do not see it—in the sense that EVERYTHING is exactly as it should be, marvelously as it should be, at every instant as it should be, luminously as it should be, without an atom of darkness anywhere: the constant Marvel, constant Light, constant Truth of everything, in the least detail, even in apparent horrors, apparent disorders, apparent falsehoods, apparent pain, apparent death, and it is just . . . SEEN WRONGLY. And the second it is seen as it is, in its truth behind the falsehood, in its bliss behind the pain, in its eternal life behind death, in its smiling and all-powerful Rhythm behind the chaos, everything is instantaneously REVERSED: there is nothing but *That*. And the death we were about to enter, the pain or despair we were about to sink into, the physical disorder, illness or murder that was about to strike us and had even laid their mortal grip upon our body instantaneously vanish, disappear, in contradiction of all physical laws, all medical laws, the millions of laws—they no longer are. And *That* shines. Instantaneously. Everything is the same. But there is no more darkness. It was seen wrongly. It was not seen divinely. Because there is only the Divine. There has never been anything but the Divine, the Joy, the Light, the all-powerful and smiling and peaceful Rhythm. And we recall Mother's words: *There is a CONSTANT Reality, a CONSTANT divine Order, and it is only because of the incapacity to perceive it that there is Disorder, the present Falsehood.* A wrong look. And She uttered these mysterious words, now illuminating: *Everything is exactly as it should be at each second—this is absolute power.* To see that is absolute power. It is the instantaneous overthrow of death and of all appearances. It is the arm of the murderer being stopped short. Because there is really only *that*. But it's not in the head that it has to be seen; it has to be seen in the body. *It's in that state of infinitesimal vibration in matter, right THERE, at that level, that the change must occur.* And She said this: *Perfection is here, always, coexisting with imperfection—perfection and imperfection are coexistent, always, and not only simultaneously but in THE SAME PLACE. Truth is here, Falsehood is here* [Mother put one hand on top of the other]; *Perfection is here, imperfection is here* [again She put one hand on top of the other]; *they are completely coexistent. The minute you perceive perfection, imperfection disappears, the Illusion disappears. The vibration of falsehood disappears ... as if it had never existed. The vibration of Truth literally CANCELS OUT the vibration of Falsehood, which doesn't exist—it only existed illusorily, for the false consciousness that is ours. . . . Which means that, AT ANY SECOND and in any circumstances, you can attain Perfection—it isn't something you must conquer little by little through increased progress. Perfection is HERE, and it's YOU who change to another state.*

And this is how the world's formidable screen of Falsehood can come undone all at once, in one second, because Perfection is here, *in the same place* and in the same terrible conditions. One changes to another state. The world changes to another state. A reversal of states. A worldwide coup d'état. There is nothing to rectify, improve, cure or transform; there is just that crust, that screen of Falsehood to get through, and it's *right there,* immediately there, the very next second. And the vibration of Truth literally cancels the vibration of Falsehood.

All at once, it is done.

There is nowhere to go to—it's *here.*

There is no "later on" to wait for—it's here.

There is no tomorrow, no maybe, no if—it's here.

But it is in matter's infinitesimal state of vibration that the vibration must change.

Then one suddenly grasps the real significance of the change of rhythm that took place in Mother's body—She was pulling on the screen of illusion in matter, the little vibration that distorts. The something in the body that contracts and says no to everything that happens. A little vibration, curled back on itself, within a closed space and defiant time. And one understands how the world can change . . . if it is capable or when it is capable of withstanding the removal of the screen without being blinded.

Not that there will be dazzling lights or dazzling apparitions, certainly not; but the world *as it is* will be such an instantaneous marvel that . . .

The miracle of the world is that it is as it is.

The illusion is in not seeing it as it is.

So we must look *there* and look at each thing as *that* miracle under veils. And who knows if our look yearning for Truth, our breathing yearning for true air, our millions of eyes yearning for a true world will not, all of a sudden, bring forth the Truth. The world as it is.

We must look at each thing as the miracle *there*.

And the miracle will be.

The great, miraculous naturalness of the divine world.

Like the axolotl suddenly transported to its true environment.

And physical life will flow uninterrupted because it will breathe true air at each instant; it will move according to the rhythm and will perceive the absolute sense of everything.

Mother's last years do not bear upon the mystery of the new species—it is settled and ineluctable—they bear upon the mystery of that worldwide "second": how to get *there*. What is the strategy?

EIGHTEEN

# The Problem of the World

That other rhythm, that time of the future species, was not trying to establish itself peacefully in a specimen free from human contingencies. She was hounded, plagued by a throng of people; her room was mobbed from morning to night. Up until She closed her door, She would receive one to two hundred people every day and continue listening to—or, rather, living, swallowing in full—their sordid, increasingly sordid, stories. This was another contradiction, which seemed to develop in the very same proportion to the inner changes taking place in her— that perfect coincidence between an exasperated Falsehood and the formidable Power one could feel suffusing the atmosphere around her was even striking, while her body grew thinner and thinner. The more She seemed to disappear, to melt away, the more the power was almost unbearable—in fact, it *was* unbearable for the Falsehood. Meanwhile, She kept dutifully reading her "oculist's chart" each day before the arrival of the mob, as if She had to continue seeing like us, and She obstinately, indomitably walked up and down her room, leaning on somebody's arm when She could no longer stand by herself— actually, She desperately struggled to keep a contact with the old human way of being. And that old way of being was constant torture for her. Despite that impossible contradiction,

She kept on going day after day, in the midst of a growing unknown, increasingly unknown and new, which created strange modifications in her: this was perhaps life, or death, or another type of life. But it was torture: *A funny kind of pain. And yet, nothing is wrong with me; the doctors say that everything is fine.* ... It was perhaps all the pain of the world. It was perhaps the mutation into the other species. It was—what?

And the process kept accelerating.

It accelerated in three ways.

## The Triple Acceleration

First, the other kind of time overcame her more and more, leaving her suspended in the middle of a gesture or a meeting. The "instant" could last 45 minutes or a second, unless it was already the next day. What time is it? She kept asking, What time is it? ... *And the funniest thing is that people think I am asleep! I am not asleep at all.* ... *A Force at work. The strange thing, in that consciousness, is the importance of one minute, which is nothing in our consciousness, while there it's important. In one minute something... general can be accomplished. Naturally, all words are stupid, but that's how it is. One minute. In one minute... To the point that the body feels that one minute like this [Mother turns two fingers slightly in one direction] means victory; one minute like that [She turns her fingers slightly in the other direction] means catastrophe. And not only for the body, but in general.* And that recalls what She said another day: *It's a key that, if you have it without being totally on the right side, it could cause a dreadful catastrophe, something like the disintegration of the world.* Is this what will permit the screen of the world to be pulled down one day, when everything is ready? Yet it is not really a "power": you *are* the entire earth, as you are your own body, without fuss or even sensation of "big" or "small"—it is simple, and terribly powerful. So one can imagine what would happen if the screen were pulled down a little too abruptly in the body. A strange condition, you must admit. A dangerous position. But completely free of all the solemn, pompous and cosmic pronouncements the mind gives out—which are nothing but mental inflations. Things are very simple *there*. They are without dimension. Simply like a body. With all sorts of ills inside. And also, symbolically, all sorts of little earthly samples around starting to think, "Well, really, Mother. . . ." *And suddenly, in the middle of all that chaos, that struggle, that friction, that suffering, and that ignorance and darkness and effort and this and that (it is far worse when it doesn't go through the mind: it's right inside the body), and it's . . . yes, truly a question of life or death in the real sense—and suddenly, just a drop . . . It isn't even a drop (it isn't liquid!), it isn't even a flash; it's . . . yes, a vibration.* ANOTHER *vibration—luminous, so marvelously soft, quiet, powerful, absolute. It's as if something were being lit up. And there's no more need of discussion or explanation or anything—you've understood: it's to become conscious of THAT, to live THAT. Just THAT, a vibration of that, and everything is understood.*

It is that other vibration She was trying to bring into the earth's body as into her own. Truly another rhythm. We call it "time" or "rhythm" or "consciousness" or "power," but it is really another way of vibrating in matter. But a way of vibrating that changes matter. Such was the deep traumatism—the traumatic bliss! It was very traumatic for those around her. *The physical world is changing,* She had said. *In fifty years people will realize it.* That new way of vibrating, which was not the mineral, vegetable or animal way, had to be implemented very quietly. Even her body had to absorb and assimilate it in a state that looked like sleep, where time "froze"—that false time of pain. And there was real life, while everything else was pure hell, more and more so. In fifty years from now, we may not not have to play the groundhog's hibernation anymore or scream with bliss (!); matter will be used to it, and it will be as natural as breathing air. But in the meantime... In the meantime, not only did She have to look after the condition of her body, the pace of the process and that strange time infusing her, but also the condition of those around her, that is, She had to plunge straight into hell to maintain a contact. She had to go slow. *It's very puzzling for all the people living with me; if I were as I should be, I think it would be rather unbearable. We must, we must endure through the transition. There has to be a transition.* The endurance was sparse around her, increasingly sparse. And this was the second of the three accelerations mentioned earlier. We could call it the "unbearable" acceleration, the acceleration of the resistance of the little earthly samples.

But there was a third acceleration, a purely physiological one, of the old physiology. There, too, an insoluble contradiction. And what is quite curious (maybe not) is that that contradiction began to manifest when Mother reached the ultimate mineral layer, the residue of the very first evolution, the last (or first) envelope of the cells—food. She could no longer eat, or less and less. A few sips of glucose or fruit juice were a kind of torture for her—it could not go down. Food is clearly the residue of the habit of the original urge to devour. Even galaxies "swallow up" one another, astronomers say. It is the sense of not having something and wanting it, lacking something and striving to take it, being separated and seeking to embrace. It is the ABC of evolution, beginning with the little protons. It actually has nothing to do with a "need to nourish oneself." It is a want seeking fulfillment, the first cry of separation. We devour because we are not *all*. For the fundamental need of evolution and of each parcel of evolution is to be all. So we "love," we take, we kill—we eat. Everything is a way of filling that primordial void, the something that set out to become a separate individual. The beginning of the cage. The beginning of death. She reached that layer in her body, and instantly the problem arose: She could no longer eat. *Food holds its own seed of death,* She noted, *and it must obviously be replaced with something else.* But what? The problem was hopeless, and it was an everyday problem. *There's this problem that, for your very survival, you are required to depend on something material, which brings back every time an old, recurring difficulty.* Every time it meant swallowing the old unconscious negation, that "atmosphere of negation behind everything," that primordial NO of an "I-me" against all the rest.... *All this is under observation right now (a very meticulous observation which might be called scientific), and, well, the cells are conscious of the divine Force and of the energy imparted to them by that Force, but they are also conscious that, in their present state, even a state of transformation, they still need that additional something from the outside—with which you swallow a new difficulty every time.... Is something that functions in the human way and is not subject to deterioration conceivable? Would this* [Mother pinched the skin of her hand], *such as it is, be capable of being transformed by the Force? Is it possible?... We'll know once it's done, but not before!* And She laughed. "But every possibility is there," I said to her. "It's just the question that matter must adjust to the infiltration of another force." *Yes, precisely!* She exclaimed. *But the point is, CAN it?* "Surely it can. If the Spirit wills it, it can. If the Spirit sees that the time has come, it can. There is no reason why not." And She laughed again. *It would be interesting to see!* She looked at the "object of experiment" quite scientifically, but the object was growing increasingly tenuous. And the more tenuous it became, the more fantastically powerful it became! Go and understand. Yet, in the midst of that power, She still needed, or felt that her body still needed, food, "to gain strength"! To gain ... That was it; She touched the crux of the problem. And it was obviously not a problem of "nourishment." It may well have been the whole problem.

A new leap had to be taken.

It had nothing to do with "stopping food"; it had to do with overcoming the NO, with conquering that layer. Or going through it to see what it hid, for, as always, the obstacle holds the key.

And the problem was compounded by the presence of her entourage. Dishes are cooked in a kitchen. And if She does not eat, why, She is going to die! No matter what happened, death was always thrown in her face, every minute of the day and under any pretext, without even thinking. It was "natural," you see. *They would like me to eat more, and I feel that eating more goes counter to the Work.... The system is beginning to refuse to work in the old way, but the doctors insist that it should work as usual—it's impossible! And that puts me in a state ... it creates a sort of conflict in the nature. You see, things are going both too fast and with a resistance of the old nature at the same time—encouraged by the doctors and habit.* This is what is constantly forgotten—the hypnotism of the physical mind. We will never realize enough how much the physical substance (including Mother's substance) is hypnotized by those do's and don'ts. Mother would struggle right to the end, and right to the end they would throw their hypnotism back at her: You must take some Coramine for your heart, you must ... And the body is like a hypnotized baby. Every second of the day, She had to undo, undo, undo that collective hypnotism—until the day She stopped struggling. And I could not understand why She did not decide to take the leap, or why Sri Aurobindo, or the Consciousness, did not tell her positively what to do. "It is better to make a mistake by listening to the new consciousness than to make a mistake by listening to the doctors!" I said to her. *But the Consciousness does not contradict anything.... I don't know how to explain.... If there were a strong and precise in-*

*dication, I would certainly listen to it, but that isn't the case....* Nothing was told to her. It was the eternal mystery. She was left to flounder all by herself between life and death—naturally, for the *body* had to find the solution! But still ... *There is the kitchen, which is in the habit of doing things in a certain way and does them in that way; there is the doctor, who said to give me such and such thing, and people listen to him; there is ... I live in such a convention that it's difficult. And always the idea that I am OLD, that I am getting OLD, and for them my consciousness must be half clouded.... Listen, they don't have faith! So I've developed the habit of saying: fine. I make myself as passive as I can—passive to the divine Will—and I pray for it to guide me. That's the only way.*

I still did not understand the full magnitude of the problem, which seemed absurd to me, even taking into account the collective hypnotism. "Why don't you use air as food?" I asked one day. "Some yogis—many yogis—have done it in the past." *Certainly not! The air is disgusting!* ... I had forgotten that She could smell the odor of an atomic explosion four thousand miles away. *They've spoiled the earth,* She exclaimed. *They've spoiled the atmosphere, they've spoiled everything! They've really made a mess of matter.... So that complicates things. Tell me, what time is it?*

It seemed as if She were reaching the heart of the contradiction in her body, with an absurd problem which looked silly to me.

And She was not being "told" anything, one way or another.

There was something to be found.

There was an ultimate resistance to transmute, one symbolically contained in a spoonful of orange juice. A central NO. There was the world at her doorstep. It was the problem of the world, not the problem of a yogic or dietetic feat of strength. And could one take that leap if no one followed? What is the point of being the new species alone?

228  THE MUTATION OF DEATH

And the triple acceleration kept accelerating.

**More and More, More and More ...**

The world, too, was accelerating toward some central contradiction, or some knot of the old story, concealed from us by mechanical appearances. Our high civilization accommodates the savage—just a few toys have changed. As what happened in Mother's body, the only merit of our terrible toys is to force us to confront the real solution, the evolutionary solution: move on to another species or die. This was the whole process. It had nothing to do with improving or idealizing the human bowl. The old evolutionary laboratory is coming to the end of its task. Mother too. It is just too bad for us if we fail to grasp the sense of it. Five years have elapsed since the rebellion of May 1968, and the suffocation is growing implacably—it will grow till we cry out once and for all, the real cry that will break the enormous Illusion. Till we land in real matter, in the real earth. The Marvel here and now, once we can no longer stand all our intelligent stupidities. 1973 is the year of the second war of Israel, the first oil embargo and the fifteenth Chinese thermonuclear bomb, the fifth French nuclear explosion, and Watergate and filth coming out everywhere, commandos and students stirring in Barcelona, Bangkok, Athens ... Almarik, Solzhenitsyn, the first American laboratory in space—toward what truer space? What less stifling air? It exudes everywhere, through all the pores of the layer of carbon, more and more, more and more, as good, as evil, as something we have yet to learn—as it does in Mother's body. This is "the transition of the earth," She said as early as 1963. *It's as if this new consciousness intensified everything in order to make it more perceptible. All the circumstances of life, the illnesses, the misunderstandings, the quarrels—everything, but everything has become very,*

*very, very acute, as if to make very sure we see them well. This is the method of the consciousness. I see very well how it works: it puts pressure so that everything that resists in the nature can come up to the surface and manifest itself, and then the stupidity or fault of that particular thing becomes evident and it must either go or else . . .* And Mother made the following strange, but very illuminating remark: *I have the feeling that, in the same way that this Consciousness sought, not positively to dissolve religions, but to enter them and remove the barriers between them, it has now decided to do the same thing with politics. It seems to be endeavoring to create, not disharmony, but a sort of—to remove the cohesiveness of people, the cohesiveness of religions, the cohesiveness of . . .* A tremendous Force, not of incoherence but of de-coherence. To take the Machine apart in its best as well as in its worst. To reach the "point of the unknown," as Mother did. And if we cannot see that, then the real meaning of our story escapes us. Are tadpoles preoccupied with improving the fishbowl or jumping overboard? We think that things are bad (they are very bad inside the bowl), but they are in fact marvelously, exactly, inexorably directed toward the exit. That exit is what we should concentrate on. *Will we have to go through a complete breakdown of the mind for people to understand? Is it going to explode with a zero at the end?*

Or will we have the courage to wage our own "immobile revolution"? To pierce a hole in that *gigantic mental bubble* and emerge into the real air? Our millions of cries are all it would take. It's right here! It's here.

*Things are difficult, grating. But that's only an appearance: it's the great Pressure of the Light—a warm, golden, powerful supramental light—more and more, more and more, more and more.*

What is going to happen?

Will this last year of Mother's body, this end of the laboratory, give us some clues to our own mystery? Or will things really have to explode for us to understand?

## Ineluctable Victory

And the little world around her was a perfect reflection of the big one, with deeper shadows and a few rare lights, as should be the case under the light source. It was the evolutionary laboratory under pressure. Everything was being played out there, on a small scale. There were no "disciples" or non-disciples, no faithful or unfaithful, no good or bad people; there were just small terrestrial samples, the old ingredients of a strange evolutionary concoction boiling furiously—would it succeed or not? It was a little like that. One sensed that everything was there. Every possibility, every marvel was there, as well as poison, Falsehood (quite abundantly!), and the old earthly calamity and clinging death, and the Miracle—if one wished. She sat in the midst of it all, so frail, so vast, so tranquil—not a shadow of a person in that little body—a prodigious and silent active witness who took in everything, looked at everything—good, evil, poison, nectar—who embraced everything in her great, immobile white Fire, without any difference, without plus or minus. It was the world, her world. These were the pieces of the problem, with each one having its own immeasurable, absolute meaning, as if that little black whirlwind that entered her room carried all the misery of the earth with it—each person was *one* misery—and that sudden, mute little flame, all the hope of the earth. That is how it was, each day like a great silent Act whose performers did not know the stake. They climbed the stairs with their dark little problem, their microscopic quarrel of a million quarrels of the world, their petty smallness of an immense smallness everywhere—but who prayed for the earth? Who had that one little cry inside for a true earth at last, that flame which wrests out the Moment and forces the door open? She was mute, vast, impassive—She was waiting. *I am millions of years old and I am waiting.* She was the prayer of the earth. *I am like a bell that nobody rings.*

Was the Moment going to be missed? As in 1950. The Moment of the earth?

Well, no. *What has to be will be, in spite or BECAUSE of everything,* She said. Oh, yes, because of all this misery, this smallness, this ugliness, because of all that rends our heart and stifles us. Something *has* to be! It has to, it has to be. It is impossible otherwise. It is monstrous otherwise. A true little prayer to unlock the Moment and open the door. For me, going to her room was like counting the heartbeats of the earth.

That so-called end is something difficult to speak of. First of all, because it is not the end; it is something quite other than what we think, even what the most enlightened think—something quite different, a mystery, really, the one I have been struggling with for the last one year and eleven months today, October 17, 1975. I am not really struggling; I am listening to Mother, but with such an intense prayer . . . No, I did not for one minute expect that "death," which most so nicely, so naturally foresaw. For me, Mother could not die. The question did not even arise. It was a sort of simple, self-evident fact—things *could not* happen as usual. She had lost the habit of dying, one could say. There was something else—but what? What was going to happen? *How* was it going to happen? was my only question. For, obviously, the transformation *was* going to take place. But how? By what way? *For me, the Victory is certain. But I don't know if it will be tomorrow or . . . I don't know what way we'll follow to get there. . . . Such an ardent faith would be needed!*

So we can only try to outline the process, return to the "laboratory notebook," walk some more in that marvelous forest where the next step was always a mystery, a suspense between the marvel and the abyss, the old earth and the unknown, impossibility and every possibility as if by magic. Impossibility—I have definitively deleted from my lexicon that word stuck between *impose* and *imposture*—forever. And when the earth consents to get rid of that impossible word, things will definitely take a turn for the better.

Meanwhile, things were strange, but they were surely going somewhere, faster and faster. I felt She was gradually settling herself in that "uninterrupted physical life," and all sorts of ailments that had constantly assailed her for so many decades were frozen, as it were, or had ceased occurring—the heart attacks, the terrible neuritis, the eye hemorrhages, the colds, the raging toothaches, the whole harvest She reaped from all the ill will around her. *I hardly know what being tired is anymore.* But there was that strange agony, without medical cause, which befell her and sometimes made her cry out in pain whenever the "outside" world drew her a bit too much into its mud or its constant invasion, into its decaying thoughts. "Mother, what will happen to sincere aspirants when You are no more?"[1] one of the little samples blithely wrote her.* That is how it was, more and more so, at every instant. What's one thought after all! But for her it was like being instantly shoved into death. *This body has become very, very sensitive. If someone comes in who is unhappy about something I did or said, it suddenly feels as if the body's nerves were being tortured. And it comes from the person who is there—who displays all the signs of devotion, etc., there are no external signs of anything,*

---

* Mother replied: "You have just voiced the thoughts of so many people. And I have only one answer to make. It is the Will of the Lord that will be realized. And the sincere people must know that [the] more they are steady and ardent in their faith, [the] more easy and quick will be the Realisation."

*no direct or spoken indication—all the nerves are tortured. . . . It's probably something that must cut the connection of the body with the Divine. It's under study.* She did many "studies." She was desperately seeking to unravel the mystery of the cells and of the cellular contagion, but *in reverse:* to pass the true vibration on to the bodies and substance around her, the vibration capable of transforming matter, the golden contagion. She was up against a ferocious, dogged resistance. All of matter resisted. But if *that* could be passed on, it would mean the end of the Screen. She was studying. She studied right to the end.

And in that strange vulnerability (her vulnerability was her means of communication!), her agony, really, which had not only to do with the world but with a deep inner alchemy taking place in her, a re-structuring or de-structuring within, resulting from the great Pressure—a whole movement She did not explain to herself, but which felt like a mass of pain inside the body, like dying at each instant to revive at the same instant— in the midst of that bizarre hell, the other state seemed progressively to grow, the other rhythm or other time which "froze" all that, or changed it into something else. And it was neither one state nor the other, but a sort of hybrid and incomprehensible existence, now painful, now blissful, now marvelous, now hellish. *It is curiously fragile at the same time; that's what is so strange. It gives the feeling of being completely outside of every ordinary law, and just . . . waiting. Like something trying to get established.* And that "something"—indeed, perhaps the state of the butterfly, the state of the next species, the incomprehensible sort of thing trying to slip through matter's meshes—was the chosen target of every assault, as if it had to be relentlessly destroyed, prevented. One cannot even say that the human samples around her were particularly baneful; they were simply the earth, the earth's condition, the earth's breathing, the earth's habit. The earth's smallness. The old catastrophic habit. *It's just a matter of overcoming all the old habit.* Not so easy. *It's like a rubber band you let go of, and everything starts again: you are in pain, you . . . But the minute the body identifies with that Vibration, it all becomes a . . . radiant expression of the Consciousness, you know, and everything is "smooth," without conflict, without difficulty. And if you can maintain that, it becomes a marvel. A marvel. Unfortunately, the influence of the outside world makes it difficult for the body to be like that ALL THE TIME, and it tends to fall back into the ordinary way. This is why that can't be established definitively. . . . Life could be so marvelously simple and beautiful! Truly, man has made it into something stupid. I very well understand that it was necessary in order to churn matter, but . . . now the moment has come for it to come to an end, for us to get out of it!* She so desperately tried to bring the moment of the other thing into this matter. And the whole difficulty, the real Screen, lay only in a nasty little vibration of the physical mind that could not keep from seeking, foreseeing, concocting its old death in every possible detail—granted, it was necessary to spur and churn this old matter, but . . . The struggle between the two states was raging in Mother's body. She was the battlefield, a "sordid battlefield" at each instant, as She sat there, pulling on the rubber band: now the marvel, now the asphyxiation, now tomorrow's earth, now the endless yesterday. *It's hard for me to keep it,* She exclaimed one day, *because every human contact brings back the old consciousness—I don't know of any people who are in that state!* Exactly. A butterfly all by itself in a world of caterpillars that kept pulling her back into the old mud. The triumphant and medical little axolotls. She had to open up the way—could that ever be made to penetrate? Her body was the site where it was trying to penetrate into the earth, to become "established," as She said.

*It's a fierce battle . . . but if I remain there [in the true Consciousness], everything is fine—everything is fine: the body is fine and everything is fine. But the moment you get out of it and participate in other movements, then you realize that everything, everything is—that it's a world of contradictions—chaos and contradictions. While there, everything is perfectly harmonious. There's a sort of demonstration of how the disequilibrium that is translated into the circumstances we call "death", (which is only a death in appearance), how both are constantly there together, as it were—the Harmony that encompasses everything, that is the essence of Life, and the fragmentation, the apparent, UNREAL division, whose existence is ARTIFICIAL and which is the cause of death—how both are intertwined in such a way that one can go from one to the other at any time and in any circumstance. And it isn't at all what people think, that something "serious" is necessary—it isn't like that; it can be the most trivial thing. It's just a difference between being here or being THERE, that's all. Being here and staying here means the end; being here and then being there [Mother made a gesture from one to the other] makes for a difficult life, a life with travails, miseries, and all sorts of things. Whereas being THERE is perpetual Life, absolute Power and ... You can't even call it "peace," you see—it's something ... immutable. But both are there at the same time: this state and that state are both there together. And man makes a kind of more or less clumsy combination of the two. Tomorrow is right in the midst of today! The other state is right in this one! It is only a minuscule distorting web that alters the Vibration as it passes. There are no centuries to travel, no miraculous transformations to bring about, no huge and endless procedures—it's* here, *the instantaneous site where life and death change into the other thing, real Life, overlife. Almost a question of position of consciousness. And the Screen falls away. That had always been there. . . . Just a few seconds of the true state in all its purity, and . . . there is a formidable power. However . . . [and this is where Mother put her finger on the problem] . . . it seems that the whole construction of the world is still an obstacle, that something is still . . .* [And

She sat looking, perhaps looking at all those little samples around her, the players in the great Act.] *And that "something" is what the consciousness is working on. Some change must take place in the terrestrial consciousness for that to become established. That's it. You can't get out all alone!*

You cannot be the next species all alone.

So what was going to happen between that increasingly aggressive negation around her (but that very negation was part of the game; it was *the* game, or terrestrial stake) and that little golden breath, increasingly gasping and fragile? And yet the victory is CERTAIN. But by what path? ... Will we know how to find the path to that ineluctable victory? It is our last path in Mother's great forest.

Where is the path? Where is the Victory, our victory? Where?

Ineluctable Victory.

## The Negation

And the contradiction kept closing in implacably.
But unquestionably in the direction of the desired Goal.
Everything is exactly as it should be.
But we have to find the meaning of it all.

We live centuries or decades with a hundred and one little meanings, and they seem to lead nowhere and just go in circles. Then comes the hour of the Meaning. We pile up discoveries that discover nothing, until the day the one thing dis-covers itself, and everything is discovered. We are reaching that hour. And the contradiction is so black only to compel us to its golden meaning. Otherwise we might as well pack our bags and go elsewhere—except that there is nowhere to go. We are perfectly trapped. Like Mother. There is no choice but to continue.

*If I live until I am a hundred, I'll have a new energy and a new life. But . . . these are just the difficult years,* She said in 1972. It was a question of time. It was a race with time. *How many years left till the centenary?* "Five years, Mother." *Five years of this hell!* . . . This was in 1973. This was the apparent contradiction, so surprising and perplexing, of that body: you sat with her, and there was that incredible, formidable cataract of power pounding and kneading you, which seemed to be capable of flattening everything (not just "seemed"; it *could*), but . . . it is as if nothing of it went into Mother's body; it simply passed through it. She was like air. "A little puppet," as She said . . . amid an incredible torrent of power. *A pipe, I am a pipe!* She exclaimed, laughing. And She poured it all onto the earth, but nothing remained in her own body. *It's very interesting. This [pointing to her body] is something apparently quite absurd, with apparent weaknesses that human beings despise and . . [laughing] extraordinary energies that human beings can't bear! It's strange.* It was very strange. *There's the simultaneous existence of unlimited power and unlimited powerlessness. And all this here, IN THE SAME PLACE [Mother placed one hand on top of the other]. I am reasonable enough not to speak, because if I told everything I see and everything that is happening and everything there is . . . people would say: That's it, she's lost her balance; with her mind she's lost her head! So I become very serious and say to myself: all right, let's take one of their so important problems—which are life-and-death problems for them —let's see, let's look at it squarely and be a little serious. . . . [And She laughed and laughed.] Well then, I still have my sanity!* Oh, that sense of humor, which saved her from the base stupidity that was heaped on her! Sometimes, though, when She was assailed by the rampant doubts about her body's weakness, She let slip a cry: *Many people have done that! They went off elsewhere, into a more or less subtle world. You see, there are millions of ways to escape—there is only one to stay, which is to have real courage and endurance, and to accept all appearances of infirmity, of impotence, of incomprehension, of . . . yes, of a negation of the Truth. But if you don't accept it, it will never change! Those who want to remain great, luminous, strong, powerful, and this and that, well, let them! They can't do anything for the earth.*

But why, why this total lack of contact with her own body? One thousandth, one millionth of *that*, of that cataract of power, could have propelled her for centuries in the blink of an eye. And I understood without understanding—without understanding the immense compassion that governs things. *If it came, it would destroy too many things,* She had already said in 1965. *When that luminous Power comes, it is so compact— so compact it gives the impression of being much heavier than matter—it is very, very, very veiled otherwise . . . unbearable.* But why, I thought innocently, could there not be just a little drop of it? And She patiently explained: *That power is REALLY all-powerful. In other words, it exists as a total and exclusive whole. It contains everything, but whatever is contrary to its vibration is compelled to change, you see, since nothing can disappear. And such an immediate, abrupt, absolute change, as it were, in the world as it is, would mean a catastrophe.* As a drop or as a whole, it is obviously the very same thing; it is withstood *only* because of our thick layer of filth.

It was enough to look at the seething little laboratory around her to understand. It seemed as if the Ashram were the center of the resistance to the Work—but of course, Sri Aurobindo had known it for fifty years! It was the symbol of all the earth's difficulties. *It is as if a superhuman power were trying to manifest through millennia of impotence. This, the body, is made up of millennia of impotence. And a superhuman power is trying, pressing to manifest itself. That's what is happening. What the result will be, I don't know. . . . I think that, in the earth's present conditions, THE result is impossible; it would be*

# THE PROBLEM OF THE WORLD

*a miracle that would upset too many things. The consequences would be worse than . . .*

So what to do if, at once, the body had to last, to gain time, and if the environment not only did not have the patience, but could not withstand that which would have permitted the change?

She was locked in an impossible contradiction. It was more and more difficult for the sips of glucose to go down. *The impression of being suspended by such a delicate thread . . . in a completely rotten atmosphere—full of incredulity, futility, ill will. A thread so delicate, it's a miracle it doesn't snap. And they don't even understand that, if that Vibration of Truth asserted itself, it would be the destruction of them, of what they think they are! The marvel—the marvel—is the infinite Compassion that holds back the destruction—it just waits. It's right here, here with all its might, its full force, and . . . it simply affirms its presence without imposing it so as to . . . minimize the damage. A marvelous Compassion. And all those idiots call it impotence!*

"Mother can't do anything anymore" was their theme, and "Mother is going fast." Some even circulated a warning: "Prepare yourself, She is about to go." An "absolutely rotten atmosphere." So?

One could almost laconically give a summary of her little cries, which were like the cries of appeal to the earth:

- It seems there is an increasingly powerful Pressure, and all the difficulties are surfacing. People are quarreling, and . . . oh! And it's all over the world.

- "She's old, She's old. . . ." That creates an atmosphere of resistance to the change. It almost creates a conflict in the being. "It's impossible, it's impossible, it's impossible . . ."— from every side.

- The entourage doesn't help. My immediate entourage has no faith.

- I am ready to fight 200 years if necessary, but the work will be done.

- I really believe that those capable of beginning the new race will be found among children. Men are ... crusted over. They are all old; I am the only one who is young! That's it, you see, that flame, that will. ... They're satisfied with small personal satisfactions that lead nowhere. While one feels that the outcome *could* be hastened if one were—if one were a conqueror! Actually, they don't care.

- It's only by desperately clinging to the Divine—but the purest and most powerful Divine—that one can avoid a ... general catastrophe. One shouldn't waste a single minute, one should cling to the Divine all the time, all the time, to compel Him to come down here. Otherwise ... So I need, I need all those who love me to understand me. We must, we must get rid of everything that still drags us down so as to be really ready to receive that divine Will. It is so frightfully urgent. Nothing else matters, nothing, absolutely nothing except a will—a will, an aspiration, an imperative need: oh, the reign of the Divine must come, it *must* come! ... I am in a hurry.

- If it went faster, it would break everything.

- Now is the time to take a definite position: that all this—disease, death, pain—all this is unreal. Now is the time.

- Sri Aurobindo said it and wrote it: "The time has come." Because He left, people thought he had made a mistake!

- There's only one direction—toward the Divine. And as you know, it is as much inside as it is outside, above as below. It is everywhere. It's in the world as it is that we must find the Divine and cling to Him—to Him alone. That's the only way.

## THE PROBLEM OF THE WORLD

- The body sees very well and very clearly the marvelous protection that is upon it, without which it would be torn to pieces.

- When I can enter my normal atmosphere, it's as if everything disappeared; I am no longer in pain. Then it comes back from the outside like a ferocious attack: people quarrel, circumstances go wrong, everything. And all that is thrown on me, so . . .

- Humanity has been waiting for this hour for centuries and centuries. The hour has come. . . . We seek a race without ego.

- To be something that doesn't add an obstacle . . . A limpid transmitter.

- Before dying, Falsehood unleashes its fury. But people only understand the lesson of catastrophe. Will it have to come for them to open their eyes to the Truth? . . . Truth alone can save us.

- I just had a fantastic vision . . . of the cradle of a future . . . not very far away. A future . . . I don't know. It's like a tremen-dous mass suspended over the earth.

- All beautiful dreams will become reality—a reality far more marvelous than anything we can possibly imagine.

- I always think of that passage from Sri Aurobindo: "God shall grow up . . ."—and you *see* God growing up in matter—" . . . while the wise men talk and sleep."[2] That's exactly it.

- There's the possibility of a fantastic success. Not in the air—here.

- There isn't any "we must wait," not any "it will come in its own time," not any—all those very reasonable things simply don't apply anymore. There is *That,* like the blade of a sword.

There is *that* in spite of all and everything—the Divine. The Divine alone. Everything else is falsehood—falsehood, falsehood, and a falsehood that must disappear. There is but *one* reality, *one* life, *one* consciousness—the Divine.

- There *will* be a miracle. But which one I don't know. It is clear, very clear that circumstances are occurring in such a way that, suddenly, things will break. But how I don't know.

- The impression is that of standing on the edge of a precipice—the least misstep could be fatal. As if Consciousness were putting a pressure on circumstances to make them more decisive.

- Even if one person can put himself faithfully at the disposal of the Truth, he can change the country and the world.[3]

- Only a violent circumstance could stop the transformation.

- There are people who are sending catastrophic suggestions. And the body is struggling, struggling to receive only those suggestions coming from the Divine. But there's still a mixture.

- I feel as if my body were as big as the world and held everything in its arms, truly as a mother holds her children. I can't explain. . . . Later.

- Everything is becoming a discomfort—a perpetual discomfort—as if my body were experiencing all the things that must disappear. And it's perpetual. Everything—things from outside, things from inside, things from so-called others, things concerning this body—everything, everything is terrible, terrible, terrible . . . As if the contradictions were accumulating in me in order for me to do the work, but I don't know who "me" is. The whole life of this body, of this poor body, is the negation of what it feels is the . . . Beauty to be realized.

- I would like to stop speaking.

- As though the battle of the world were waged in my consciousness. And it isn't just of one person; it's the subconscious of the earth. It's endless. Yet it must be done. . . . To stop it means to stop the work. To continue it means it will take . . . I don't know . . . it's endless. It's as if this consciousness were the point of junction and implementation. So the only thing I can do is remain quiet, very quiet: to let the divine rays pass. It's the only solution. The Divine is the One who must . . . do battle.

- A bliss that's right here, ready for us.

- The slightest contradiction that enters the atmosphere makes me feel so uncomfortable that I feel I won't be able to stand it. . . . It's . . . I don't know how it is. . . . It's like a negation, a painful negation.

- The ultimate outcome is evident.

- I feel like screaming. . . . When I am still, I have almost unlimited power. And when I am in my body, I feel so uncomfortable. . . .

- It's become so acute. And at the same time with the knowledge: This is the time to win the Victory. Something coming from above, like this: hold on, hold on—this is the moment to win the Victory.

What could possibly happen?
What would be the last path to that ultimate, "evident" outcome, that ineluctable Victory in spite of everything, or because of everything? Oh indeed, because of everything! "The reign of the Divine must come, it *must*—it must! . . ."
Or else, what? Begin all over again?

## NINETEEN

# The Impossible Solution

She was so vast and so perfectly still in the midst of the great battle being waged. One merged into her as into an immensity of soft snow, and yet so fiercely burning in its stillness. One went far, far, and forever—and yet it was close at hand. One was at home, as if within one's deepest inner sanctum, and yet the pulse of the world was right there. One was bathed in love, yet it was merciless war ... in utter silence, as if it were beyond all wars, as if they had been won forever. And the little drops of her words, her little breaths from across eternity bore a fire, spoke of hell or the Marvel, of the contradiction and the eternal question, in such a completely even tone of voice, as one who watches a river flow: here it turns to the right; here it turns to the left. It was transparent truth, without color, without ripple—pure. Impersonal. And yet it was She.

And more tenderly so when She laughed.

### On the Threshold of a Great Secret

True, She did not laugh much anymore. It was getting very close. But She never gave the impression it was close. The seriousness of the situation was never apparent, except by the fact that She was increasingly absorbed in that strange time. There was such a bath of compact and light eternity around her that it was impossible to imagine where and how it could possibly end. Clearly, death could not exist there. But there were all those looks of death around her. That was the assault. Her 95 years could have been just as easily 395 years, but time counted for the others. One did not even feel that her body was getting older—"old" was such a strange word near her—but all those looks . . . And I recall these words of hers with a flash of comprehension: *I have the feeling that the visible form is as much (at least as much) the result of how one is perceived by others as of how one is oneself. I don't know how to explain it. . . . When somebody else sees you, you see yourself the way he sees you. But there is a certain way of being, produced by the true consciousness and felt in a very concrete way, which is . . . not exactly in contradiction, but altogether different from the way people see you. . . . That's why there is something to find that can be independent of everybody's influence.* We may never know how miraculously supple and fluid matter, bodily matter, is, but it is if coagulated, hypnotized by a habit. Truly, the world is completely distorted. *Matter* is completely distorted. It can be miraculously otherwise . . . if we get out of the hypnosis—if we see otherwise. If we are otherwise. Mother was seen as old and dying. That was the real problem. And She was not fighting for herself; She was fighting to free a parcel of matter from that dreadful hypnosis. Oh, how they believe in death! *The power of death is that they all want to die!* She exclaimed one day.

But there was the "other way," "produced by the true consciousness," that "son of the cells," slowly developed and amalgamated by the prayer of the body, the aspiration of the body, the Mantra repeated millions of time night and day. She had seen that new body a second time, in 1972: *I don't know whether it is the supramental body or a transitional body, but I had a completely new body, in the sense that it was asexual.... It was very slender, very pretty. Truly a harmonious form. What was very different was the torso—the breathing. It had broad shoulders. That was important. The chest, however, was neither masculine nor feminine. And as far as the stomach, abdomen and all that are concerned, there was just a contour, a very slender and harmonious form which was certainly not used in the way we use our body.... Evidently, what will be very different—what had become very important—was the breathing. That's what that being mostly depended on.* And I said to Mother, "I've several times had the impression that, rather than a transformation, there will be a concretization of the other body?" *Yes? But how?* "Well, of course, we don't know the details of the transition, but instead of this body becoming the other, the other one would replace this one." *Yes, but how?... Of course, if the body I was the night before last were to materialize... But how?... What is the process of the transition? The transition that for most people is like going from the waking consciousness to the sleeping consciousness and from the sleeping consciousness to the waking consciousness. It's still a step. It's still doing this and that [Mother twisted two fingers to indicate a shift or change of consciousness from one state to the other].... We know nothing. It's peculiar how we know nothing!*

For her, there was no longer "transition," no longer a waking side and a sleeping side, a side of death and a side of the so-called living. It was all one. That was her strange, paradoxical life, as in two worlds at once (two worlds for us, for our distorted vision in the cage). She was building that overlife in which there are no more "sides." *I see many scenes of nature, like fields, gardens, but everything is behind nets!* She said, laughing. And those "nets" were so visibly and manifestly (and symbolically) the web separating us from the "other side." I could almost see them as Mother spoke. I saw sorts of delicate sardine nets swaying in the breeze, between her and a true and so very smiling earth. *There is a net of one color, of another color ... And this has a meaning. Everything, absolutely everything is behind a net. We are—it's as if we moved around with nets! But it's not just one net; it depends: in its form and color, the net depends on what's behind it. And that's ... the means of communication! You understand? Fortunately I don't speak to anybody, otherwise they'd say I'm cracked! And all this with eyes wide open, during the day, if you can imagine! For example, I see my room (say I am here with a visitor) and AT THE SAME TIME, I see one landscape or another, and it changes and moves, but with a net between me and the landscapes!* All this is very lovely, and we can certainly conceive of an overlife ... without a safety net (!), but in the meantime there is still an enormous Screen, with a perhaps very slender body in true matter, but also another one in this matter, which is poorly regarded and driven to decomposition. So? ... So what to do? The day the Screen falls away will be a happy day indeed (unless it is another sort of catastrophe!), but we feel that in order for it to fall away, the present matter just has to open itself, take the first steps in its own flesh, extricate itself from that kind of leaden fixity or general hypnosis. This is precisely what was painfully taking place in Mother's body in the midst of the enormous collective Negation. That kind of "old rag," as She referred to her body, was the site of the experiment, the site of the transition, one could say—if it did not happen there, where would it happen? Through what crack in universal matter? The question was not to leave the old rag to go play around in a less wretched story—it was taking place *there,* in those so poorly regarded 95 years. And if nobody around her wanted to have anything to do with it, who in this damned collective matter would? The laboratory was perfectly representative of the whole. It was the "battle of the world." Hence Mother was bumping against the eternal mystery: "It's peculiar how we know nothing." And time was pressing: *It has to go fast.*

And sometimes, She sat looking in front of her, her right hand pressed against her lips, and one felt such an intensity of question. She looked at the conditions, the implacable conditions—and what way out? It was not the way out for her body! It was the way out for the earth. But *who* wanted it? Who? Where was the pure little flame in all that mire which came up to her room dressed in white, with its official yogic smile—there were others, of course, those nobody speaks of, those who have no name, no title, who toiled obscurely, for love, washing dishes or greasing cars, those who had such a lovely little light in their eyes when you chanced upon them. Those were the ones for whom Mother kept holding on—the ones who almost never saw her, who could not even bring her the little breath of their pure love. But there was all that dark carapace around her—it was *in that* She had to work; it was from *that* Negation She had to wrest a cry of consent. And She toiled and toiled. And sometimes it seemed hopeless. *If only there was a certainty. If, for example, Sri Aurobindo said: That's how it is. Then it would be very easy. But what is difficult is . . . You see, you are surrounded by people who think you're sick, and who treat you like a sick person; you are surrounded by a certitude that you are fast approaching the end. So this poor body is struggling with all that, and it doesn't know. It doesn't concern itself with it, but it doesn't have any certainty about the final issue. So the only thing it can do is be quiet, have faith, and . . . endure.*

Of course Sri Aurobindo did not say anything! The body had to find the solution. For the body, to find is to do. And yet, invisibly, one sensed that She was on the verge of something, in spite or because of that very Contradiction. One could feel her more and more rapidly alternating between two extremes of splendor and disintegration, of victory and abyss, as if the two were perhaps going to merge into some incredible other thing.

Oh, it felt so near! The more ferocious it was, the more obvious and palpable it was. And Mother knew nothing; She was learning nonexistence in order to be able to exist and to last in that frightful swinging from one side to the other. She built the bridge between the two. She *was* the bridge. Was it death for good? Was it the other thing? *The apprenticeship of personal nonexistence. To forget the Divine even for one minute becomes a catastrophe. From time to time, for a few seconds, true blissful consciousness—but only from time to time and for a few seconds. That's all. The rest of the time, battle. Once, twice, for a few seconds: Aah! ... And it's gone. Does ... this body have to be left and another one built? I don't know. ... It doesn't correspond ... I have never been told such should be the case.* "But what would we do here if you left?" I protested. "It seems as if the only moments we can breathe are when we're with you!" *But it doesn't wish to leave. It doesn't know. But something has to be done for this body to be more plastic so it can be transformed, otherwise, well, it will be for another life. Though I must say that ... Sri Aurobindo himself told me, "To have to start all this again, the whole childhood and all that unconsciousness—no." Before leaving, He said, "No, I'll come back when it can be done in a supramental body." But there have to be some people who can last indefinitely. I have the feeling that that's possible. I can remain like this for hours, in a sort of receptive contemplation, and it seems like a second. Time is certainly curious ... there's no more sense of time. I feel ... I feel ... I am on the threshold of a great Secret ... but ... It isn't mental, not with thoughts. It's ... "something."*

## The Great Immobility

We do not really know that great secret, at least not yet. It cannot be something confined in a mental formula—perhaps a new way of functioning in matter. What can be the great secret of the caterpillar seeking to become a butterfly, unaware of what a butterfly is, and wrapping itself up in a sort of death? She did not know, She spoke less and less, and speaking was very difficult: even our conversations were eavesdropped on. She did not have a single refuge left, except for that growing "something" in which She tried to take me along, silently, while holding my hand. She was building one last line of communication. She wanted to make me feel the place of the experience. And suddenly, one day, She abruptly came out of the experience, as if compelled by something imperative: *I want to see you every day.* One of the attendants was called in, the schedule was arranged.... And it lasted once ... after which it was "not possible"—there were so many reasons for it to be "not possible." The wall of Negation was slowly, inexorably closing in on her. *I have no more control. I've lost the habit of saying "I want."*

Something was ineluctably pushing her toward an impossible point—or perhaps the very site of the solution?

And simultaneously, "something" seemed to be developing—was it the same something? The other face of the Negation? The lever within the obstacle? We never understand anything of the world because we always see "adverse" things. But truly, everything is a mystery. We will only understand when we reach the end. *That sense of time is what I don't understand. ... Oh, I feel, I know—I know without any doubt that my body is being accustomed to something else.* But what? It seemed as if everything revolved around that question of time. *Life is a torture if I am not exclusively turned to the Divine. It's the only remedy. Otherwise life is a torture. Existence becomes intolerable. The only remedy is to be ... where time doesn't exist.* What was happening in her body?

The laboratory notebook is increasingly slim and laconic: *I am walking a very thin and narrow line.* Every minute was a dizzying balance between an indefinable "something," the "marvel," as She called it, perhaps the state of the next species, and the death of the old species. A sort of living death. *There's like a Pressure—a formidable Pressure—to make the necessary progress. I feel it in myself, in my own body. . . . And it was really the body of the whole world. . . . But my body isn't afraid; it says: all right, if it's over, it's over. That's how it is at every minute: the True Thing . . . [Mother lowered her fist] or the end.* What was so extraordinary in Mother was that mixture of imperturbable peace, so vast, so removed from everything, and at the same time that terrible, almost ferocious will amid utter calm. I have seen many things in my life, but sometimes I shuddered in Mother's room. She was like a living sword of light. *. . . Each minute is imperative: life or death. Not the approximation that has lasted indefinitely. For centuries we haven't felt positively bad, and we haven't felt positively good—it's no longer that way. The body knows that the formation of the supramental body requires everything to be ENTIRELY under the Influence of the Divine—no compromise, no approximation, no "it will come"—only this [Mother lowered her fist again], a terrible Will. . . . But it's formidable because there's constant danger. You see, I don't know, perhaps a hundred times a day, the sensation (I mean, for the cells) is: life or disintegration. . . . And one wonders if it was not life and disintegration. This was truly the impossible paradox. . . . And if the cells don't contract as they usually do, everything is perfectly fine. It's as if, out of a sort of obligation, the body were learning eternity.*

Every detour always seemed to lead back to that other time, as if the key lay there.

Then one last blow fell on her. It was in April 1973. What Mother called the "transfers." One after the other every body function, every organ underwent a "change of authority," namely, a transition from nature's old, automatic functioning to the conscious functioning of the great Consciousness—the supramental functioning, what Mother and Sri Aurobindo called "willing automatism." The uncoupling from nature's old laws and the coupling to the other rhythm. And one morning, visibly shaken, Mother simply said, *It's my nervous system that's being transferred to the Supramental. . . . I have the feeling—it's worse than dying.* It was the last transfer. In other words, nothing in her body was governed by the old law anymore. Perhaps the last connection or ultimate transition to the other state. And one can well imagine how positively intolerable it must have been to go through all that while living, speaking to people, signing checks, swallowing a thousand poisons from every side. A caterpillar builds a cocoon for itself before undergoing the process. *But I think . . . I think I can transmit the divine Vibration. So if you want to stay here . . .* And Mother held my hand to take me into the experience. Perhaps She wanted me also to feel the key. "Why, to be with you is like being under a cataract!" I exclaimed. "It feels like a fire of purification, like—it widens you, fills you up—that is IT. It is since you are apparently powerless that I've begun to feel . . ." *Listen,* She interrupted, *I've accepted. The Lord asked me if I wanted to "undergo the transformation," and I said yes—I would have said yes in any event. But it's—for the ordinary human consciousness, I'm becoming crazy.*

Yet at the same time—at the same time—there was such a new, such a strange phenomenon beginning to manifest in her body. . . . Perhaps it was really the "dangerous unknown"? Dangerous, because we do not know what it is, because it is new. Would a caterpillar not find the butterfly state dangerous? Happily, it is not aware of it! But Mother was alive to every minute of it. *It's so utterly different that one wonders—I sometimes wonder how it's possible. There are times when it's so new and unexpected . . . it's almost painful.* What was happening? I so much wanted to understand, to feel. . . . Meanwhile, the "unexpected" inexplicably grew and developed. There was visibly a sort of acceleration—toward what? And one morning in April 1973, as I was immersed with her in the experience for half an hour or so, She suddenly opened her eyes: *Why do I feel like screaming?* I was stunned. I immediately thought that, perhaps, I had brought in some poison with me—oh, we are so full of dark little woes. "Well, I wonder, perhaps I'm the one who's causing you pain?" *No, my child, I am that way ALL THE TIME—it's not you at all. It's something.... It isn't at all painful, but it's something....* And She sat looking out toward the big yellow copper-pod tree, looking at the something that made her want to scream, and that was not painful.... *Personally, I think—I think it's something so new the body is frightened. That's the only reason I can see. I begin screaming, and... it's of no use. The only thing to do is stop and change.* Perhaps this was the change taking place.... *Yes, that must be it: something so new that the body... doesn't know how to take it.* And She turned to me to try to understand what was happening: *You didn't feel something while we were meditating? What do YOU feel?* And it was so hard to tell. "Something like a fire merging with *your* Fire," I answered. *But what do you feel?* "I don't know ... a great Power." And She nodded, insisting as if I had not felt the main thing: *You don't have some kind of perception?* So I finally said, "No, Mother, what I primarily feel is that great flame disappearing into you, and then a sort of vast immobility—a powerful immobility." And suddenly, I remembered that "immobility of fire" Sri Aurobindo spoke of. *Ah, that's it!...* And Mother smiled as if She had put her finger on the something I did not understand. *Ah, that must be it! Yes. The body must get frightened. Yes, that must be it.*

And what is "it"?

As for me, at the end of the meditation, after that torrent of power which seemed to churn me and swallow me up, I felt the same thing almost every time, as after a cataclysm: something stretching far, far out and becoming immobilized; nothing stirred anymore, not a breath, as if the body had stopped breathing or breathed without breathing. Immobile. A compact, burning immobility, and yet without any weight. Everything stops. Everything had stopped.* Yes, perhaps that "eternity of the body." But what was it *for her*? For her body? I felt only a reflection, a minute reflection of what She felt. So? And She said, "That's it, that must be it. . . ." An immobility like that of death, felt as death by the body, but a death . . . that would be alive. Words are absurd, but it is something, a state evidently quite outside of life, of the breath of life, and yet breathing, yet not death—another, very new way of breathing. The death of the old species, of the old breath, the birth of—of what? And it is really like death. "The body doesn't know how to take it." It feels like screaming, yet it isn't "painful." Is this really what the transformation is? The transition? The zero hour? Obviously, you cannot be a butterfly while remaining a caterpillar. But how is it possible with eyes wide open, with fifty people waiting at your door and a few others watching your every move? Is it possible to grow the other thing inside the old body? How is the physiological transition made? She was visibly at the very threshold of the unknown, of the great secret.

And the only answer was that immobility in another kind of time.

The great immobility.

---

* Let us point out that this immobility has nothing to do with the one experienced at the opposite end, when consciousness dissolves up above in the supracosmic pallors and the body sinks into a trance of oblivion. Indeed, here, in this physical, or rather physiological, immobility, it is the entire bodily agitation that is stilled, but it is coupled with a sort of enhanced keenness of perception that is quite the opposite of sleep or trance. Everything is perceived—people, things, clocks, thoughts, the slightest vibrations—and from very far away, in a sort of physical extension of oneself, which elects, or does not elect, to know what is there. But the ticking of a clock in New York City was as distinct as the one next door. One is right in matter. But a complete matter.

Was it death? Was it life?

Was it transformation or disintegration? "It's almost the same process!" She used to say. *As if to show you that in order to conquer death, you must be prepared to undergo death.*

So?

It seemed as though She were being trained to enter death alive.

## As Inside an Egg

Where was it all leading?

To me, it was clearly a question of time and patience. "Given time, everything would change," She had said. There was no need to undergo death, I thought, except perhaps as a radical step lasting a few days or weeks, and everything would then be changed. I felt that that immobility without time was a sort of laboratory condition to allow the change to take place in the very core of the substance, that "infinitesimal vibration" in matter, the intraatomic movement that causes the original hardening, the coagulation of the Screen. This was the "missing side" of the atom, the one that had the capacity to modify the movement by "freezing" it in its lightning-fast immobility. I did not really know, nor did Mother. *You see, we don't have the knowledge, the slightest knowledge of what supramental life is. Hence, we don't know if this [Mother pinched the skin of her hand] can change enough to adjust to it or not. And to tell the truth, I am not worried about it; it's not something that preoccupies me much. . . .* And I looked at Mother uncomprehendingly. *. . . Because what preoccupies me is building that supramental consciousness so IT becomes the being. It is that consciousness that must become the being. That's important—the rest, we'll see. It's as if you were preoccupied with wearing one set of clothes rather than another—that's the equivalent. . . .*

I did not understand what She meant. . . . *And in order to do that, all the consciousness that is in the cells must combine, organize itself and form a conscious, INDEPENDENT being—the consciousness in the cells must combine, organize and form a conscious being that can be at once conscious of matter and of the Supramental. That's it. That's what is now taking place. How far it will go, I don't know.* But that appearance, the "set of clothes," kept dominating my mind, as if it were the most difficult thing to change and transform, while every time Mother seemed to see it as an ultimate consequence without any difficulty whatever: "At the end, it will be nothing; a mere breath and it will be done. It's everything else that's difficult." Of course, I knew that that "son of the cells," that new body formed with the combined consciousness of the cells, was there, already formed, and this is what really preoccupied Mother: that it become an independent entity—that is, independent of the body—capable of being simultaneously conscious of our old material world and of the world of true matter, the supramental world. But what did that mean? If this old set of clothes is nonetheless the site of the ultimate transformation, the place where the Screen falls away, the bridge to true matter, then what is the role of that cellular body vis-à-vis the old body? What will open the door to the other type of matter if not old matter itself? What will bring in the other substance if not the old one? The supramental world will not fall from the sky or break through old matter's doors without something *on this side* lending itself to the process. *One* receiving body, one place for it to happen is needed, no? . . .

Unless the Screen is worn down everywhere at once, the darkness purged everywhere at once, the layer of carbon so thoroughly and unbearably asphyxiates itself that the new body will suddenly break through all the pores of our despair— perhaps that is what will happen. "There *will* be a miracle," She said.

But while waiting for that last "breath and it will be done," there is this transition, this time to be gained, and what is the role of the cellular body, which seemed to be Mother's first preoccupation, the body that is "to become the being," that is, in some way replace or relieve the old body? Does it mean that Mother was about to let go of the old body? . . . But that cannot be true! Mother had said and repeated time and again, *Death isn't AT ALL a solution! There's no solution, except if this [Mother touched her body] is transformed.* She had said it again in 1971. And so many times: *I always, always receive the same answer, which is not an answer in words but the answer of a Knowledge (how shall I put it?) . . . a Knowledge of FACTS: it isn't a solution. Something that comes from a very absolute realm and makes you sense or understand or grasp the uselessness of death. Consequently, we are seeking another solution. There MUST be another one.* And She was categorical: *Death is the acknowledgment of defeat, so . . . To me, it's a falsehood. Death and falsehood are two of a kind. It's still the memory of a disastrous past. And there is no point in abdicating because it must be started again the next time around. . . . To start again in a little baby?* And Mother shook her head.

There is only one way out—through the supreme Door.

So She was searching for another solution. Or rather, She was fabricating the other solution. What was the actual role of that cellular body if it was not simply meant to replace the other one? . . . I do not know if there is an answer to that question, or if there will be one before we reach the end of the evolutionary process (or at least this part), but my recurring impression is that the cellular body is the one capable of assuming all the old body's functions, of maintaining all the connections and contacts with the old world of matter as we know it, of *physically* surviving death without its making any difference to our central consciousness ("It would make more difference to them than to me," She said), and hence of keeping the old body (truly like a custodian and a support) as long as it is necessary for it to be transformed and recast in its image. A sort of luminous cocoon that safeguards and protects and regulates the deep, intraatomic transmutations of the old substance. More and more, Mother enveloped herself in that body without time and wear and tear, which could keep her going indefinitely in its conscious automatism—provided the rest of the world allowed it. And indeed, one could feel that, as long as She was there, She was completely outside of the 70 or 80 pulse beats, and of the 95 years that tomorrow will irrevocably make 96. She was outside of all medical and baptismal conventions . . . provided the rest did not keep throwing its old mortal convention in her face. An "independent" entity, a tiny golden vibration that repeats and repeats itself in the depths of the cells, and that will keep repeating itself until the process is done. Until the golden invasion occurs in every point of the old body. It was just a question of time and patience. *I have the feeling that if I last till I am a hundred [She added in that same conversation of 1972 in which She referred to the formation of an independent body as her foremost preoccupation], in other words, six more years, a lot of things will be done, a lot—something important and decisive will be accomplished.*

And again in 1972, one day She made this very revealing remark (all the more revealing that it concerned the mind of the cells): *Only when the Supramental manifests in the body-mind is its presence permanent.* And since I did not understand what had to be "done" (we always feel we have to "do" something!) in order to drive the Mantra into the body's cells, that is, make the Supramental permanent in the body, Mother replied to me, *I don't know what has to be "done," because it's spontaneous. Perhaps this is the way: a contemplation of the Divine. That's what the natural state should be. Truly, I think there's the same sense of powerlessness as a newborn baby. You understand? It's even curious, but the feeling of the body is that of being completely enveloped, like a baby in a blanket. Exactly like that. Two or three days ago I felt something pressing on my heart—it was painful. It was painful; I really felt that . . . the body felt it was the end. But then, immediately, I felt myself as if enveloped, like a baby, in the arms of the Divine. And after some time (which seemed long), when it was fully in the Presence, the pain went away. The body did not even ask for it to go away—it just went. Absolutely the impression of a baby enveloped in the arms of the Divine. Extraordinary. I think that . . .* [and this really opens up new perspectives] *. . . I think it is now excessively sensitive and it has to be protected from all incoming things—as if it had to work within, you know, as inside an egg.* She sat looking for a while. *Yes, that's it. That's exactly it. I think a lot of work is being done inside. Oh, in terms of the old way of being, it is more and more stupid! But the new way is beginning to take shape. Oh, how one wishes one could be enveloped like that, for a long, long, long time! . . . It's coming. We must be patient.*

As inside an egg.

*And the material consciousness is repeating, OM Namo Bhagavate. . . . It's like the background of everything. OM Namo Bhagavate. . . . You know, a background that is a material support. OM Namo Bhagavate. . . .*

She was spinning her cocoon of light.

It was March 1973.

That "new body" is truly the cocoon of the body in transformation.


## Sleeping Beauty

In my heart I could feel the Moment was nearing—really the Moment of the earth, that for which we had so much struggled, suffered throughout all these ages. For me, there was no doubt: we were approaching—but by what path? I had the feeling of watching a poignant Performance next to which Aeschylus and the mystery plays were pale copies. Now, it was the Performance of the earth. And She was there, smiling, immobile, as if wrapped in white light. *Time is no longer the same. . . . And I can no longer eat. Well. What is going to happen, I don't know.* "Very good things, surely!" *You're sweet*, She replied. "But I mean it, Mother. I am sure!" *Yes, of course, so am I!* And She laughed, took the big white hibiscus that I handed to her and put it on her knees: *What is it?* "It's a 'Grace,'" I replied. *Then it's for you. That . . .* And She remained motionless, her hands open on her knees. *Can't speak anymore, can't eat anymore . . . and time passes like lightning.*

Food, I thought, was her next step. It was the natural "seed of death," the very symbol of the first self-consumption, Love's devouring copy. The last change of functioning leading—where? It was also the last communion with the old earthly way, and this, I sensed, involved far more than just another little change: that absurd symbol meant confronting the whole world, as if crossing that line was to violate the human law itself, setting out on an irreversible course—and people around her struggled to make her eat. If She did not eat, "She was going to die," of course, you see. And She yielded to their hypnotism, sometimes even telling them: Force me to eat, even if I don't want to. It was so pathetic to see that struggle of the body to escape, then slip back into, the suggestion, and to escape again. "She was going to die" was written on every wall around her, whispered everywhere, and it kept battering and pounding her cells. "The structure of the world is still an obstacle," She said. Nobody believed in the miracle! Nobody wanted to believe in "the other way." It was hard for one body to believe against the whole world. Mother was nothing but one single body, nothing but an earthly physiology desperately trying to effect the transition of the species—but what do you do when the species does not want to, does not understand? And, in effect, good will was as harmful as ill will—a different will was needed! *I am in the battle night and day. . . . It is not for ONE body; it is for the earth. This body has become a sort of representative, symbolic object—it's the battlefield. It's the battlefield, it's the field of victory; it's Defeat, it's Triumph, it's everything.*

To violate the law all alone?

One sensed that that step hid something very critical. A very critical change was needed in the terrestrial consciousness, in that little laboratory around her.

Was the Hour going to be missed?

And sometimes, they poured so much defeat, impossibility and negation on her that her body no longer knew: wasn't it mad to fight single-handedly against all those scientists of the earth? Those irrefutable little axolotls. *I don't know. I don't know what will happen! There are times when it's so difficult, I wonder if the body can hold on. But I would like . . . Oh, did I tell you, yesterday or the day before, all of a sudden, for two or three minutes, my body was seized by the horror of . . . The idea of being put like that into a tomb was so dreadful! That I couldn't have stood for more than a few minutes. It was dreadful. And it's not that I was being buried alive, but my body was conscious! It was "dead," as people say, because the heart was not beating—but it was conscious. That . . . was a dreadful experience. All the "signs of death" were there, meaning the heart was not working, nothing was working—but I was conscious. It was conscious. They should . . . they should be warned not to be in a hurry to . . . If the time of transformation comes, if my body grows cold, they shouldn't be in a hurry to put it into a hole. Because it may be . . . it may be a passing phase. You understand? It may be momentary. You understand? You understand what I'm saying? . . .* Oh, I understood so very well! And that day, I recalled a question I had asked Mother a few years earlier, when I wanted to know if one could "have the experience of death without dying." *Absolutely!* She had said. *One can have the experience yogically; one can even have it materially provided . . . [She burst out laughing] provided the death is short enough so the doctors don't have time to declare you dead!* So now I saw the scenario. What if they declared her dead? For one HAS to die, you see. That is the earthly law. *That's how it is. Come on, don't be foolish.* She, too, saw the scenario. *. . . I feel,* She continued, *that an effort is underway to transform the body—it feels it and is willing, but I don't know if it's able to do it. You understand? So for a while it could give the impression that it's over, and it would be only a passing phase. It would start up again . . . it could start up again. Because I might not be capable of talking and saying it at that time. So I am telling you. . . . I don't know. But I would like someone to prevent that blunder from happening, because all the work would be lost. Some people with authority should say: YOU MUST NOT, Mother DOES NOT WANT. . . . You could . . .* "Who will listen to me?" I interjected. "They will say I am crazy. They won't even let be come near your room!" And I did not know how prophetic my words were. Then She added, *It sounds silly to make an issue out of it. It's better to say nothing.*

One "makes an issue out of it," and it is only the issue of life on earth.

It is silly indeed, just business as usual.

It is difficult to make History alone.

Was it the sign that it was "not possible"?

And weeks passed "like lightning" (for her). *I am nothing but a force pressing against a world of obstacles,* She had said twelve years earlier. And it was like yesterday. *Any effort to maintain the old way has become—produces a discomfort, an almost unbearable discomfort.* She could no longer remain in the old way; something had to happen. *We just have to hold on, that's all!* And something happened. One day—one does not know whether it was black, because each night conceals a

## THE IMPOSSIBLE SOLUTION   263

greater light—things took a turn. . . . In what direction we do not know. It was probably still *the* Direction, because there is only one. On that day—so dark—it is as if Mother had touched the key. It was April 7, 1973.

She did not smile that day. She was grave, "as when I pull the weight of the world." She had a white lotus on her knees. *I seem to encompass all the resistances of the world. They come to me one after another, and if I weren't . . . if for one minute I stop calling the Divine, I am overcome by an unbearable pain, my child. To the point that I now hesitate to mention "transformation" to anyone, because if that's what it is, well, you really have to be a hero. . . . You see, something in the body would scream all the time. Yet, I have the feeling there's something so simple to do for everything to be fine. . . . But I don't know what it is. It's strange, I say to myself: Does the Lord want me to leave? And I am . . . quite willing. Or does He want me to stay? . . . There's no answer. And that's . . .* Yes, always that silence, perhaps till the very end—not A SINGLE answer. Why? If we could understand why, we would have the key to what happened. . . . *I truly have the impression there's something to do, and everything would be absolutely fine—but I don't know what it is.* "I have the feeling you are absorbed in an increasingly rapid movement," I said. *Yes, that's true. . . . You see, I have ONE solution for transforming the body, but it's. . . . It's never happened before, and it's so . . . incredible. I can't, I can't believe that's it. But for me that's the only solution. . . .* I was wide-eyed. . . . *The body feels like going to sleep ("going to sleep" in a certain way: I would remain totally conscious) and wake up only after it is transformed. . . .* Suddenly, it sprang out like a revelation: Sleeping Beauty! Yes, that's it—cataleptic trance, the cocoon of transformation. It was obvious. It was the only solution. The "so simple" thing to do. Then She immediately corrected herself, . . . *But people will never have the patience required to go through that, to take care of me. It's a colossal task, a Herculean work. They are nice, but they're already doing the maximum. And I can't ask more of them.... That's the thing. It's the only thing to which the consciousness says: Yes, that's IT. ... But who? Who? It's almost impossible to ask that of those people who take care of me.* "They'll understand—at least some will understand," I innocently replied. *But I can't tell them.* "Well, *I* can," I foolishly retorted. *Will they believe you?* I was dumbfounded. For me, it was so obvious. *You might perhaps explain to them in front of me ... when they come,* Mother continued. Then She closed her eyes, plunged into the experience, taking me along in that strange and powerful immobility. And this is where She suddenly came out of her state, saying, "I want to see you every day." That "every day" lasted only once.

Then it was time.

They came into the room.

And my heart sinks.

One of the samples approached, one of the "guards." And there was an immediate explosion. Thirty years of safety valve suddenly burst. Oh, I will not repeat, I cannot repeat that horrible monologue, that outburst of anger. And She, so white and immobile. Oh, no one is to be blamed; it was not a human being who stood there, and I am not sure if one person on earth could have borne that formidable Pressure day after day—it was the earth saying NO. These were the laboratory conditions. This was where She had to work. There were no "good" ones or "bad" ones, just the samples of the terrestrial enterprise. *What is closest to the center of descent is very much churned,* She had already said nine years earlier. But these are the facts, and I cannot erase them from the Story. She tried to speak: *I can't speak ...* "Don't speak, Mother." *I would like to explain ...* "I am not interested." *It's because there's an attempt to transform the body.* "When it happens, we shall see." *But you don't want to know?* asked Mother in her little childlike voice. "No, I don't want." And that was that.

That day destiny was sealed.

The solution was there, together with the impossibility of the solution.

She could no longer continue, and the only door was being closed on her.

She could not transform herself without flattening everything around her that resisted.

Like Sri Aurobindo.

There only remained the "hole."

And yet, yet I felt, I knew, that the ultimate Solution lay in that very impossibility—the solution nobody had thought of. The very reason why She was not told anything. The last path to ineluctable victory.

There is not any NO anywhere; there is a supreme YES, everywhere and forever, which follows imprescriptible ways, through all the no's, all the refusals, all the good, the bad, the violent, the peaceful. And everyone does exactly what is required ... without knowing it.

Humanly, we grieve (oh!), we lament, we allow ourselves an outburst of anger or reprobation. And it does not make any difference; it is all the same—we understand nothing. There is "something" that follows its imperturbable course. In the ultimate negation is hidden the supreme Door.

I had a white lotus in my hands when I came out of Mother's room, and I did not understand anything anymore. I only understood that an invisible page had been turned. "One day, they will close Mother's door on me," I said mechanically to the person who was with me. I did not know I was being prophetic for the second time.

The enraged voice could still be heard: "In thirty years I've seen enough humbug!"

*There's something so simple to do....*

If the earth only knew how simple the thing to do is for everything to change.

So simple.
A mere little "yes" in the heart.
A yes that even melts tombs.
Then the Hour would come.
A mere breath, and it will done.

* <br> * *

Mother had seven months left.
The last path.
Sleeping Beauty needs a Prince Charming.

TWENTY

# The Last Path

What would the path be now?
Surely, there had to be another path; the terrestrial enterprise could not fail. Perhaps She will tell me what that path is, for I do not know. She talked to me for so long; surely, She will tell me the Truth again. I do not need soothing tales; I need an earth of truth.

Death was not a solution. Nor did I once think about it from that black 7th of April until that 17th of November, which fell on me like a cry: NO! It cannot be. That's not it—that's not how it is! How I cried no that day, and how I have cried no since! It could not be. Or what?

For the last one year and eleven months, I have not stopped tracking down that "what," listening to it, struggling with it and pursuing it in each and every line of this book, as if it were the very blood of the earth trying to know, to understand. I set every possible trap to capture it, refusing to delude myself with anything, probing every corner of that terrible and marvelous forest, and at the end I am simply left with a prayer: The truth, the truth, let the truth be.

All we want to know is how to build that true earth.
What is the path? The last path?
And I seem to see her smiling.

## The Last Meeting

So let us follow that road again. Seven more months. I did not know that destiny had taken a turn on that 7th of April. I looked to the future without alarm, with just a great question in my heart. I could see Mother's increasing pain, but . . . dare I say that it looked to me like a battle against unreality? One just had to come into contact with her atmosphere to understand, to feel, far more solidly than any pain, that formidable Reality. She struggled with the earth's past. She struggled against all the ghosts of evolution, symbolized by all that negation around her. It was the earth's No that had to be overcome, not "death"—that ghost of ghosts. Mother was *not* 95 years old! She was as old or as young as one wished.

And what did they wish around her?

What was the reality for them?

*I have the feeling of being pulled in opposite directions by the old world and the new one.* . . . The laboratory notebook is getting very sparse; I had no more questions in my heart. The day after April 7, the daily stream of one to two hundred people was stopped. Only the dozen or so "regular" disciples remained, but they were the dangerous nucleus—the ones who *knew* She was going to "die." Oh, they knew everything; they were well-informed!—it is not the 150 others who were the problem. She had indeed to continue the battle to the very end. *I feel like screaming. . . . So . . . I haven't seen anybody this morning, and they're all waiting at the door. What shall I do, my child?* And that "What shall I do?" was so poignant. It was truly "What shall I do?" What? What *could* be done? "We need you, Mother," I said. And that was all. *Oh! . . . Oh, thank you*, She said. And it was so heart-breaking. Then She plunged into her eternity as into a second, and half an hour passed. Whereupon they would come near her and ring the bell: *They're relentless.*

Then, on April 14, the last operation, which I mentioned earlier, occurred—the transfer of the nervous system. From then on, the whole body functioned automatically, independently. It was the little vibration of the cells, the new body, that enveloped it, made it move, breathe, feel. Nothing was controlled by the old laws anymore. There was obviously a logic in all this, but the logic of what? It is quite probable that the caterpillar's last operation in its cocoon is precisely that effacement of the nervous system. But in this case, without a material cocoon to protect her against the assault of the environment—or, rather, with a cocoon of light that was forever punctured by outside contacts, which caused her pain every time—what could possibly happen, since the cataleptic trance was impossible and her entourage would never have the necessary patience? What was the solution? "What shall I do?" *I have constantly to hold onto myself not to scream,* She told me on May 14. *And from time to time, there is a marvelous moment—but it's brief.* She could not sever the contact with the outside world; that contact was precisely what permitted the transformation of the old matter—otherwise one simply disconnects and it's over. She could not even exteriorize herself outside of this sorry earthly mess, as She had done for so many decades, in a fraction of a second. *For so many years, so many years, I just got into my bed, and aaah, I merged with the Lord. But now, it's forbidden. And that's what is most painful. The moment I start "going out," I instantly feel a terrible discomfort—it's NO. And if I persist, I literally start screaming as if I were being tortured. . . . It's only when I am concentrated here that it gets better.* And that "better" was a strange hell. Undoubtedly, She had to find the solution in the body; She was stuck there until She found it. And the only possible solution was transformation—yet circumstances seemed to want to prevent the only condition that would enable her to undergo the operation. So? . . .

So the 19th of May came. It was my last meeting with her, but I did not know it.

I went up the small stairway carpeted with wool. The door opened, and there was that ray of light on the nape of her neck. Her armchair was turned toward Sri Aurobindo's tomb. She was bent over herself, in her immobility, her hands open on her knees, as if offering all this world, all this misery: "What You will, what You will...." She was truly the earth's prayer. Then She opened her eyes and smiled, took my flowers: *What is it?* "It's Joy in the Physical." It was a tiny, champagne-colored hibiscus with a fire-red heart. *We really need it!* She exclaimed. *And you, no questions?* And that day, I, who had remained silent for weeks, was full of questions. "Actually, I thought of a text by Sri Aurobindo, in *Savitri*, in which He clearly says, 'Almighty powers are shut in Nature's cells.' "[1] *Ah? that's interesting. . . . He doesn't say anything else?* "No, not on this subject. It would seem that the consciousness of your cells has awakened, but not the power." *Did you say "awakened"?* "Yes, because if the Power were awakened, there would be no weakness in your body." *No.* And I was really so stupid not to understand that if that Power "awoke," manifested itself, it would be shattering, the destruction of all the resistances around. There also She was stymied, stymied on every side: the very conditions of the transformation could not occur without "a miracle that would upset too many things." So? . . . "But what has to be done to awaken that?" I asked. *Faith, our faith. Knowing it and having faith. . . . But, you see, my physical, my body, is very rapidly deteriorating—what could keep it from deteriorating?* "I do NOT think it is deterioration," I said. "I don't think that's what it is. I have the impression that you are physically being led to a point of such a total powerlessness that the most total Power will have to come out." *Ah, that's it*, She said. "The Power will be COMPELLED to come out."

And She remained silent, one hand pressed against her lips. Crows cawed in the big yellow copper-pod tree. It was full of flowers, a golden cataract. It was May. *Or else I can . . . I can leave this body, no?* And She said this so matter-of-factly, perhaps to tease me, or to tease the earth a little, just to see if it cared or not. "Certainly not, Mother! Now is the time to do it. It has to be now. I really feel this is NOT disintegration, not at all. It is NOT a disintegration. Personally, I have always seen that, in the face of the opposite extreme, the other pole comes out. Likewise, the supreme Power has got to come out of that sort of apparent impotence. This isn't disintegration in the least." *For me, it boils down to a question of food. I have more and more trouble eating. Can this body live without food?* "But Mother," I insisted (and I had the feeling of fighting with her, or perhaps fighting against the suggestion weighing down on her, that "completely rotten" atmosphere), "I really feel that you are being led to the point where something else will have to manifest itself. You see, so long as it hasn't reached the point of . . . impossibility . . ." *Ah, it's almost reached that point.* "Yes, Mother, yes! That's what I feel. I feel you are reaching that point, and something else will emerge. . . ." And She did not answer anything. ". . . This is not at all the end. On the contrary, it will soon be the beginning." *I was told the beginning would occur when I am 100 years old, but that's a long way off!* "No, I don't think it'll be that long. I don't think so. I don't think so. Another functioning is going to take over. But first, the end of the old has to be reached, naturally, and that end is the terrible part." *Oh,* She said, *really, I don't want to say anything; I don't want to insist, but . . . really . . .* And She shook her head, and there was all the pain in the world in that shake. *The consciousness is clearer, stronger than it has ever been, yet I look like . . .* "Yes, Mother, you ARE going to emerge into something else. I feel it. This is not faith speaking; it's something else in me that knows. You are going to emerge. It's really like something telling me: that's HOW IT IS."

She closed her eyes on her eternity.

I never saw those eyes again.

Those were my last words.

Two days later, everybody was barred from entering her room.

It was the end of all communication with the outside.

"There will be no more communication," I said to an astounded disciple. They did not understand that by closing her door to me they had cut the thread! Oh, there are always two ways of looking at things, and everything is decreed, everything follows the divine Plan. She probably had to be left alone. But sometimes one wonders if things could not have been different. Yes, they could have been! They could have been.... But the whole earth should have been different. So? ...

There was nobody left to pull on the thread.

She was alone with the Negation.

Or perhaps alone with the Answer.

*It isn't to rest that I aspire, but to Your integral Victory,* She said.

By what path?

And I am reminded of Sri Aurobindo's words, the last lines He dictated in 1950, "The Book of Fate" in *Savitri:*

> *A day may come when she must stand unhelped*
> *On a dangerous brink of the world's doom and hers.*
> *Carrying the world's future on her lonely breast,*
> *Carrying the human hope in a heart left sole*
> *To conquer or fail on a last desperate verge.*
> *Alone with death and close to extinction's edge,*
> *Her single greatness in that last dire scene,*
> *She must cross alone a perilous bridge in Time*

*And reach an apex of world-destiny*
*Where all is won or all is lost for man.*[2]

## Hold On

What happened during those six months?

All appearances wanted death, all doors were closed; there was no way out—except the supreme Door. Except the best possible path to ineluctable Victory. It is my theory, my stubborn faith. And I know I am right. But . . . I am not seeking relief. I do not need to be relieved; I have a Fire inside that laughs at all death. And that's how it is. And Mother is with me. But the earth? But that beloved, wretched, painful earth—what is its path? How does Mother's last path join with the earth's path? The earth was her path; She cut the path for the earth. What is the path? *Sri Aurobindo told us in a categorical and definite way that the supramental creation would follow this one. Therefore, everything the future holds in store for us has to be the circumstances necessary for its advent, whatever they may be.*

It is this "whatever they may be" that we would like to decipher, without nonsense, without sentimentality—the truth. Unadulterated.

But the truth of the earth.

I can only tell my experience. There must have been a reason for her to build that last line of communication.

Here are the facts:

I was therefore waiting for the last "change in functioning": the cessation of food. For me, it was a simple and obvious consequence of the past six months. She appeared one more time on her balcony, on August 15, 1973, for Sri Aurobindo's birthday. She was completely doubled over, and still fighting. She was such a tiny figure up there, on that big poop deck, while the peanut sellers clanked their metal pushcarts as at a fair. It was such a small, very small world, habitual. Were we going to get out of that habit? Truly, if the end result is only a "superman," it is hopeless—something else is needed. A kind of physiological catastrophe is needed for something to change—not a war, not millions of bombs that change nothing, and we make little babies and we start over again. NO. Something else. How I pursued that "something else" in that bent little figure. Then She grasped the railing with both hands and, slowly, very slowly, She disappeared in her little golden cape.

I would only see a body, three months later. But I could not believe it. Can one believe in the failure of the earth? Could man fail after the monkey, or the frog after the tadpole? It is only a question of how. That's all. There must have been more than one monkey to say no to man. And so what?

On November 10, She began to choke, as if She could no longer breathe—slow asphyxiation. The time of that kind of breathing was coming to an end. The time of that kind of air. What did death matter to her! One cuts the thread, as simple as that, and that's the end of the sack of misery—She could have cut it some ten years earlier, thus avoiding a lot of misery. She never cut it. Even "dead" on that chaise longue, She seemed to be bursting with consciousness—a fierce consciousness. What happened during those days, those last days when She was choking? "I am not told anything...." She was not told anything up to the very end. She had to remain there to the end, till the last second. She had to enter death fully alive. That was the horrible thing. One day in 1972, She had told me the following—three things at once which seemed unconnected, three things I have already mentioned separately, but She had said them together: *The body, the body's prayer, when it became conscious of what was happening, has been this: Inform me when the moment of dissolution comes, if dissolution is necessary, so that everything in me may accept that dissolution, but*

*ONLY IN THIS CASE.* She was not "informed." Nothing in that body had accepted dissolution—She was fiercely present on that chaise longue, to the "end." Then, in that same conversation, She also said, *My body is LIVING THROUGH THE PROCESS. And it's only when I am still, in a sort of contemplation of the cells, that . . . everything is fine. Time disappears—everything, but everything changes into something else.* . . . She lived through the process till the end. It was merely the continuation of the process. And that same day, on the same subject, She also said, *It's become very acute. And at the same time with the knowledge: "Now is the time to win the Victory," like that, coming from above: "Hold on, hold on, now is the time to win the Victory." It's really interesting.* All three things at once. Putting all three things together, what do we get? . . . Probably what She experienced more and more acutely until that 17th of November. She was not told anything; hence She simply kept going. The process kept going. "And She went out like a candle," reports one of those who watched over her—no heart attack or anything of the kind. Not even any trance or coma. She kept going as long as She could. She had to remain there fully till the very end. Why? . . . It probably means that the process had to continue in death. She was not allowed—her entourage did not allow—the cataleptic trance; they would not have had the necessary patience. But they allowed death; that was "recognized," accepted as a medical reality. And since She could not perform the operation on her bed, She went into the tomb to do it. There, nobody would disturb her. It was the best possible cocoon. They would even burn incense sticks over it. *In order to overcome death, you must be prepared to undergo death*, She had said. In a tomb, there is no trick. Trance was still a yogic trick. There is no more door. The only way out is the supreme Door.

On November 14, at midnight, She asked to walk: *I want to walk, otherwise I'll become paralyzed.* She held on to the arm of one of the attendants and walked ... until She turned blue.

And the days grew more and more painful. She refused to eat, then accepted, then refused again. She was scolded, coaxed like a patient. She was always half-seated, her back covered with burning bed sores. Every twenty minutes, She would ask to be lifted from her chaise longue: *Lift me up.* ... On the night of November 16, She again asked to walk: *I want to walk*—oh, She wanted to enter death on her two feet. It was denied to her. She fought till the very end, as long as there was a breath left in that body. On the afternoon of the 17th, the signs of choking grew worse. At 7:10 P.M. her doctor massaged her heart. At 7:25 P.M. her breathing stopped.

And that is when the incredible madness began.

Barely seven hours later, at 2:30 in the morning, they moved her out of her atmosphere, took her downstairs to the "Meditation Hall," laid her on a chaise longue under searing lights, and turned her over to the crowd. Oh, who is to be blamed! They did what is always done in "such cases." They did what is "customary." And there were three medical doctors to certify her death and ... And, look, you are not going to make a fuss over this, are you? I could still hear Mother's words: "They must not, they must not, Mother does not want. You will tell them. ..." And yet those people must have had some sort of yogic knowledge, no?

I arrived in the middle of that crowd around five in the morning, informed by rumor. She lay there, so thin, on her chaise longue, in a white satin dress. With that ... almost ferocious concentration on her face. Nothing like Sri Aurobindo's smile in his massive peace. A relentless concentration, as in the days when She was in the thick of the battle. And all that conscious body, bursting with consciousness—it was palpable—

offered up to those thousands of vibrations of grief and death and...

Hold on.

And the fans roared. They roared for two days, at 85 degrees Fahrenheit, under a suffocating zinc-plated ceiling ablaze with golden neon lights, while the whole town filed past her—truly the best way to decompose a body. And She watched all that. At times, I almost expected her to give one of those little starts, as when She came out of her concentration: What time is it?

I had no pain in my heart. I was like a stone. I stared and stared at all that, at that incredible spectacle.

Hold on....

Then, at 8:15 A.M. on November 20, they put her into the box. I was standing to the right of that coffin. She was half-seated on white cushions, her hands on her knees. There was a ray of sunlight on the nape of her neck. Then the cover was lowered; there was no more ray, nothing. There were half a dozen men driving 25 screws into the wood.

They took her away.

A solemn voice was delivering solemnities into a microphone on the Ashram's roof. And it went on and on. People were full of sorrow, full of thoughts. I could see through them all, through everything; the world was transparent. It was a dreadful farce. A Lie. An enormous Lie, a drama of death such as we see in dreams. It was not true. There was not one minute of truth in it all. That death was a dreadful lie.

She was lowered into the tomb.

As soon as it was over and I could decently take leave of that solemn masquerade, I fled.

I am sorry, but that is what I saw and felt. It was my experience of those two days: an unreal Lie.

So what does it all mean?

What is the meaning?

## The Most Beautiful Fairy Tale

And here is the last fact.

On November 18, in the middle of that incredible masquerade, while I was in that crowd, staring and staring uncomprehendingly at that white little form amid the roaring of fans, I had the most powerful experience of my life. I was incapable of having any experience. I was like a stunned rock with a splitting headache. I just stared, without even a prayer in my heart, nothing. Had She suddenly gotten up and walked out of that unbelievable commotion, it would have seemed to me perfectly sensible. She did not get up, but all of a sudden I was seized by something which literally pulled me above that headache and that dreamlike crowd, and—it was like an all-engulfing flood. I knew what Power was; after all, Mother had not held my hand and taken me into the experience for nothing. But here it was not a person having an "experience"; it took place outside of me. I was nobody; I was merely witnessing something. I was immersed in a tremendous flood of Power, made of elation— maybe love, but it was an elation that was love—elation like a torrent, without letup or slackening, and it kept ringing— a tremendous peal resounding over the universe. All the floodgates were wide open. And it spoke; it pealed words in my ears as well as over the whole world—a formidable but soundless voice: NO OBSTACLE ... NOTHING STANDS IN THE WAY ... NO OBSTACLE ... NOTHING STANDS IN THE WAY ... And it kept ringing and ringing, each word reverberating as if all the bells in the world were ringing together in a tremendous peal of bronze: NO OBSTACLE ... NOTHING STANDS IN THE WAY, NO OBSTACLE ... And with such joy, such triumph, oh, something so bursting with delicious but irresistible laughter, washing everything away, toppling the walls, bursting open the gates—nothing stands in the way ... no obstacle. As imperative as a Last Judgment. A cataclysm of joy.

I held out for a quarter of an hour, then I went out into the street lest I could not contain myself. And it still kept ringing. I walked to the sea, my body shaking. Finally it quieted down. And there was no "Mother" in this or any "me," or even an experience—unless the world itself was having the experience. Yes, in fact, it was like the first manifestation of "something" over the world. We can put labels on it, but it does not care a damn about labels. It was a formidable Fact. Something happened on that 18th of November.

Perhaps the first terrestrial wave of the joy of the new world.

So all the wailing, all the admirable close disciples who cried so much after having consigned her to death, all those faces with their mask of dignity, and questions of money behind, questions of prestige behind, questions of—oh, they were full of questions; they were about to assume control of affairs—it was all so ridiculous. This was really the masquerade. And Mother laughed. She rang in our ears, swept that ashram away like a straw in the wind, along with all their silly little walls and all the sainthood they were about to saddle her with—NO OBSTACLE... NOTHING STANDS IN THE WAY... NO OBSTACLE...

And they said, "Oh, the transformation is stopped; oh, the Hour of God is missed; oh, the Moment is past; the world wasn't ready"—the fools! Could the pithecantropus have ever stopped the torrent of mental evolution and avoided becoming men?! Did they really think Mother so small as to care about the four walls of an ashram or about their mortal conventions? She was there, before my eyes, bursting with consciousness. She was going to enter that tomb since it was the only place where they could tolerate her for any length of time—now they have all the necessary patience. But She was laughing.

This is all I know. That peal of joy over the world.

That is the Fact.

So what does it all mean?

We are truly reaching the end of Mother's forest; I know nothing more. It has been one year and eleven months since they put her in there, alive. For the last year and eleven months I have kept repeating, "What? What does it mean?" It is my own bell. What does it all mean? What is there? What is really happening? What is the truth? . . . And I recall Sri Aurobindo's words:

> *In that tremendous silence lone and lost*
> *Of a deciding hour in the world's fate, . . .*
> *Alone with her self and death and destiny*
> *As on some verge between Time and Timelessness*
> *When being must end or life rebuild its base,*
> *Alone she must conquer or alone must fall.*[3]

Now, She is in that "tremendous silence," in that night, that cocoon of gray marble. Cement above, cement to the right, to the left, night, concrete slabs—a tremendous silence. Each of her cells repeats and repeats the Mantra, endlessly, like a golden little pulsation. She is undergoing the formidable operation. She is rebuilding the base of life. The "process" continues. This is what She had been prepared for for months: "My body is being accustomed to something else"; this is why She was not told anything, because She had to enter there alive—I still hear her little cry the first time She had the vision of her death. *Nothing, absolutely nothing works in the usual way anymore! The body can no longer eat, no longer . . . And the Consciousness that is devoted to helping it in the work made ab-so-lu-te-ly clear to it that leaving is not a solution. Even if before there was some curiosity to know what will be, that curiosity is gone. As for the desire to stay, it's been gone for a long time. And any possible desire to leave when things get a bit . . . stifling has gone with the idea that it won't change anything. So there's only one thing left for the body to do: to perfect its acceptance. That's all. The only thing that comforts it (and not for long) is the idea that: what you are doing is useful to all; you are not doing this for you, a small silly person, but so the whole creation benefits from it. . . . I don't know, I don't know what's going to happen. But I'd like . . . I'd like not to be put into a box and stuffed into . . . because it will know it, it will feel it, and that will only add one more misery to all those it's already had. I am telling you this so you can tell others if necessary. . . .* [I was informed nine hours after Mother's departure.] *. . . It doesn't desire it, it doesn't fear it—it will be as it should be, that's all. However, it really would like people to understand . . . to appreciate the effort it has made and not decide to box it up and throw dirt on it. Because, even after the doctors have declared it dead, it will be conscious. The cells are conscious. That's all I have to say.*

She is in there, alive.

Aeschylus and Orpheus look pale.

There is really nobody to blame for this incredible tragedy; each of the actors probably did exactly what was required. I recall Mother telling me, in 1969, the astonishing circumstances of a young Ashram girl's "accidental" death, as though everything had been orchestrated down to the last detail and conspired to force her to die. It seemed as if each person made just the right gesture, had the right distraction, the right lapse, the right three-minute delay. And Mother knew this young girl had to die, that she *wanted* to die, that her soul had arranged all the circumstances to "facilitate" her departure. *When you see from the point of view of that Consciousness, there's such an amazing perfection in the organization of things that it's almost . . . frightening. All our emotions, our reactions, all that seems absolutely childish. We know nothing, my child! Day after day, day after day, I am more and more convinced: we know nothing. And we think we know, we think—we know nothing. We are before hidden marvels that completely escape us because we are* stupid. That's what Sri Aurobindo wrote in Savitri: *God grows up on earth—God GROWS UP—while men . . . [She laughed] . . . while wise men talk and sleep. And the work will go unnoticed until it is completely finished.* And that's how it is.

What is concealed behind that spectacular performance of Mother's "end," *the Eternal's dreadful strategy*,[4] as Sri Aurobindo said? What marvels? What lost silence? Or what?

What stratagem?

And I still hear Mother's words: *Seeing the world as it is and as it irreparably seems to have to remain, the human intellect has decreed that this universe had to be a mistake of God. . . . But the supreme Lord replies that the comedy is not completely over, and He adds: Wait for the last act.*

What is that last act, that last recess of Mother's forest? That last path of the earth?

And I hear my bells again.

NO OBSTACLE, NOTHING STANDS IN THE WAY . . .

And that "We know nothing."

We know nothing. What fine words are we going to come up with—but it is not words that the world needs! It has been saturated with every conceivable theory. But if it could hear those bells, if those bells started to ring on the earth?

Will they ring?

We can say: "To be sure, one day Mother will emerge radiant from that tomb, and the transformation will be done." And that may be true. But Mother has never been inclined to dazzle people. A great spectacle is not what will change the earth. What is needed is a profound miracle, a change in this substance, despite anything we may think of it, say of it, want or not want, believe or not believe. What is needed is for this life—this physiological life—to establish a new base for itself. We have to be seized by the miracle, in the miracle, immersed in it. We have to get out of the layer of carbon, shatter it; we have to get out of this silly little axolotlian brain with its ridiculous little panaceas, and of this small, so small, humanity. What instrument will perform the miracle? What lever? What open sesame in this unyielding substance?

And I seem to see this:

A division. A fantastic division.

An old, outmoded, doddering and incorrigible humanity—the one that is showing more and more signs of ruin, that is merely false matter held together by a necktie or by a garment of white cotton and a frozen yogic smile. That one is dying all by itself. It is heading faster and faster toward terminal dissolution. It is in the throes of final suffocation. No need to push it—it is in the process of falling to dust, solemnly. And then...

> *When darkness deepens strangling the earth's breast*
> *And man's corporeal mind is the only lamp,*[5]

... something very young, very new, bumping about right and left, unknowing. The new species—not even aware it is the new species, except that it wants a different air. And...

And this is where the miracle *can* occur.

We do have the open sesame.

The whole work, the real work of Mother and Sri Aurobindo, was to open up the consciousness of the cells, to open up that fortress. The shattering of the old genetic code: the old way of looking at things, of understanding, of feeling—of dying. The cave of the axolotls. A new little vibration repeating and repeating itself in the cells. We must capture that little vibration the way others, at the beginning of this wretched story, captured the mental vibration. And it is not complicated; there is not far to go, no superbrain to build, no cross-legged meditations—a Mantra. A password, whatever it is, but it has to be the cry of our being, the breath of our breath amid this universal decomposition, something that enables us to break through. The last buoy. And we cling to it. And we repeat and repeat the Mantra until it breaks through the crust of banality, of everyday triteness, the millions and millions of useless things we live through in the name of something that never comes. We repeat, obstinately, like a mule, until the cells capture the vibration of that call—then they repeat it night and day, automatically, stupidly ... and marvelously. This is where the Marvel begins. This is where the miracle begins. The cellular, physiological miracle. Because Sri Aurobindo and Mother opened the way. It is not that we have to cut through impenetrable layers—the way is open. And so everything is possible. It is a world where everything is possible—it was only impossible because we thought it impossible. There is no need to believe, you see; we just have to go *there,* to reach *that* level. We have to touch that. And the marvel is that, when we touch it, the new world begins to grow *by itself,* without any need to will or make an effort or understand on our part. It works by itself, automatically, spontaneously. It develops by itself. It alters the movement of things, we do not know how; it makes us do certain things, we do not know how; it brings out all sorts of meanings where there was only stupefying blackness; it connects and interconnects all sorts of unexpected paths—there is a tremendous complicity in everything. And then a power ... a strange power, in no way powerful, in no way towering, but so fantastically magical, as if it dissolved all obstacles, all shadows, all fears, all illnesses—a rout of ghosts in every direction. But ... one position to hold firm to: *that.* Always, in every circumstance: That, the Mantra, the new Truth, the supreme new Possibility. We just hold on to that no matter what.

Then, not dozens, but thousands, and perhaps hundreds of thousands, perhaps even a few millions, of young shoots in the world have understood and seized the lever, and they pull on that, yearn for that, call that. ... Why, there's instant multiplication! A gigantic contagion, as if just a little bit of naked, experienced, prayed-for Possibility aroused thousands and thousands of other Possibilities—compelled the Possible. And everything begins to change physically and materially. The old laws come undone. We look at all this, and it melts away; it ceases to be. It no longer is. It is finished. And the more we laugh to see that, to touch that, to experience that, the more it grows and develops with a marvelous smile, as if it wanted only to smile, to smile forever, to smile everywhere. The end of the mental Ghost.

And we are there.

The Screen falls.

And Mother is there, and the real earth is there.

The end of the axolotls.

The real earth is only in our stupid tomb, our impossible tomb—the one in which Mother was buried because nobody wanted to believe it possible.

Then She will not have to "get out," to dazzle people, because everything will be a Miracle and everything will be a Marvel. She is only on the other side of *our* Screen of impossibility. There is no miracle to perform; it is already done. There is no Marvel to invent; it is already invented. The Marvel is everywhere. The Miracle of the earth is *right there*. We have to go *there,* to the real level of the Marvel.

And perhaps death will affect those who do not know how to see the Marvel. They will suffocate all by themselves, inside their tomb in the open air.

And the division will be made automatically.

The dead go toward death, and the living toward life.

A little golden vibration within the body's cells.

An open sesame of the real earth.

A tremendous peal reverberating over the ruins of the mind.

No obstacle, nothing stands in the way.

Mother is not dead! She is right there, alive, laughing. She is waiting for us to get out of our stupidity—it is not She who needs to get out! It's we.

And when the crust of carbon is worn thin enough, even her body will come and laugh with us. She is wearing down Death from the inside. In truth, her body only went into the tomb to wear down our own death. It is her last sacrifice to our impossible Falsehood. She is waiting for us to consent to see WHAT IS. She is wearing down Unreality.

This is Sleeping Beauty.

*She lived in spite of death, she conquered still.*[6]

Sleeping Beauty needs a Prince Charming. Perhaps she needs many Prince Charmings.... For her Prince Charming is the earth's soul in the shackles of death.

She is slowly effectuating the mutation of death in her cells. Each little cell like a cry of appeal to the truth of the earth—there are one hundred billion cells in a body. And when we have had enough of the masquerade, when our cry joins with hers, then the masks will fall.

It will be "the end of Death."[7]

What if we helped a little?

The mutation of death is now.

*Well, this is how it is. You see life, you see what it is, you are used to such an existence, and it's dull and grim (some people enjoy themselves, but it's because they enjoy themselves with very little), well, there is a fairy tale behind it. Something that is about to take place and that will be so very beautiful, beautiful beyond all expression. And we will participate in it. You don't know, you think that when you die you'll forget and leave everything behind, but it isn't true! And all those who are interested in a progressive, joyous, luminous, beautiful life, well, they will all participate in it in one way or another. Now you are not aware of it—in some time you will be. There you are. Yes, a beautiful story. And Sri Aurobindo was trying to pull that story down onto the earth, and it's sure to come. . . . And if you wish, you too can pull to help it come on the earth. . . .*

What if we pulled a little?

A little golden vibration in the cells.

A radiant new species.

A body of our joy.

*And it's true! It's the most beautiful fairy tale in the world. No other story is more beautiful than this one—I am going to tell you the most beautiful story in the world . . .*

> There you are, Mother, your geographer has finished his map. May the real earth be.
>
> Deer House
> Nandanam
> October 26, 1975

# WORKS OF THE MOTHER

| | |
|---|---:|
| Ascension vers la Vérité | 1956 |
| Belles histoires | 1946 |
| Un Centre International Universitaire | 1952 |
| Commentaires sur la Dhammapada | 1960 |
| La découverte suprême, 1912 | 1937 |
| Éducation | 1952 |
| Entretiens 1929 | 1933 |
| Entretiens 1930-31, Aphorismes et Paradoxes | 1957 |
| Entretiens 1950-51 | 1967 |
| Entretiens 1953 | 1975 |
| Entretiens 1956 | 1968 |
| Entretiens 1957 | 1969 |
| Entretiens 1958 | 1972 |
| Le grand secret | 1954 |
| Paroles d'autrefois | 1946 |
| Prières et méditations, 1912-1919 | 1932 |
| Les quatre austérités et les quatre libérations | 1953 |
| Quelques paroles | 1951 |
| Quelques réponses | 1964 |
| Sri Aurobindo, Pensées et Aphorismes, vol. 1 | 1974 |
| commentés par la Mère vol. 2 | 1976 |
| Vers l'avenir | 1949 |
| Whites Roses (original English) | 1964-70 |

# WORKS OF SRI AUROBINDO

## 1. Indian Tradition

| | |
|---|---|
| *The Foundations of Indian Culture*, 'Arya' Dec. 1918 - Jan. 1921 (New York) | 1st ed. 1953 |
| *On the Veda*, 'Arya' Aug. 1914 - Jan 1920 | 1st ed. 1956 |
| *Hymns to the Mystic Fire* | 1st ed. 1946 |
| *Isha Upanishad* (translation & commentaries) 'Arya' Aug. 1914 - May 1915 | 1st ed. 1921 |
| *Eight Upanishads* (translation & introduction) | 1st ed. 1953 |
| *Essays on the Gita*, 'Arya' Aug. 1916 - July 1920 | 1st ed. 1922 |
| *The Renaissance in India*, 'Arya' Aug. 1918 - Nov. 1918 | 1st ed. 1920 |
| *The Significance of Indian Art*, 'Arya' 1918 - 1921 | 1st ed. 1947 |

## 2. Philosophy - Sociology

| | |
|---|---|
| *The Life Divine*, 'Arya' Aug. 1914 - Jan. 1919 | 1st ed. 1939 |
| *Ideals and Progress*, 'Arya' 1915 - 1916 | 1st ed. 1920 |
| *The Superman*, 'Arya' March 1915 - Aug. 1915 | 1st ed. 1920 |
| *Thoughts and Glimpses*, 'Arya' 1915 - 1917 | 1st ed. 1920 |
| *Thoughts and Aphorisms* | 1st ed. 1958 |
| *The Hour of God* | 1st ed. 1959 |
| *Evolution*, 'Arya' 1915 - 1918 | 1st ed. 1920 |
| *Heraclitus*, 'Arya' Dec. 1916 - June 1917 | 1st ed. 1941 |
| *The Supramental Manifestation upon Earth* 'Bulletin' 1949 - 1950 | 1st ed. 1952 |
| *The Problem of Rebirth*, 'Arya' Nov. 1915 - Jan. 1921 | 1st ed. 1952 |
| *The Human Cycle*, 'Arya' Aug. 1916 - July 1918 | 1st ed. 1949 |

*The Ideal of Human Unity*, 'Arya'  1st ed. 1919
  Sept. 1915 - July 1918  Revised 1950
*On the War*, 1940 - 1943  1st ed. 1944
*War and Self Determination*, 1916 - 1920  1st ed. 1920
*Man—Slave or Free?* 'Karmayogin' 1909 - 1910  1st ed. 1922

## 3. Yoga

*Elements of Yoga*, 1933 - 1936  1st ed. 1953
*Lights on Yoga*  1st ed. 1935
*More Lights on Yoga*  1st ed. 1948
*Sri Aurobindo on Himself and the Mother*  1st ed. 1953
*The Mother*  1st ed. 1928
*The Yoga and Its Objects*  1st ed. 1921
*The Synthesis of Yoga*, 'Arya' Aug. 1914 - Jan. 1921  1st ed. 1948
*Letters*, 2 volumes (On Yoga I & II)  1st ed. 1958
*The Riddle of this World*  1st ed. 1933
*Bases of Yoga*  1st ed. 1936
*Correspondence with Nirodbaran*  vol. I  1st ed. 1954
  vol. II  1st ed. 1961
*Letters* (translated from the Bengali)  1st ed. 1961

## 4. Literature - Poetry - Drama

*Views and Reviews*, 'Arya' 1914 - 1920  1st ed. 1941
*Letters*, third series  1st ed. 1949
*Life - Literature - Yoga*  1st ed. 1952
*Conversations of the Dead*, 1909 - 1910  1st ed. 1951
*The Phantom Hour* (short story), 1910 - 1912  1st ed. 1951
*Kalidasa*, 2 volumes, 1893 - 1905 (Baroda)  1st ed. 1929
*Vyasa and Valmiki*, 1893 - 1905 (Baroda)  1st ed. 1956
*The Future Poetry*, 'Arya' Dec. 1917 - July 1920  1st ed. 1953
*Collected Poems and Plays*, 2 volumes  1st ed. 1920
*Poems Past and Present*  1st ed. 1946

| | |
|---|---|
| *Poems from Bengali*, 1893 - 1905 (translation) | 1st ed. 1956 |
| *Savitri* | 1st ed. 1950 |
| *Last Poems*, 1937 - 1944 | 1st ed. 1952 |
| *More Poems* | 1st ed. 1957 |
| *Vikramorvasie*, 1903 - 1904 (Baroda) | 1st ed. 1911 |
| *Songs of Vidyapati*, 1893 - 1905 (Baroda) | 1st ed. 1956 |
| *Rodogune*, 1893 - 1905 (Baroda) | 1st ed. 1958 |
| *Ilion* | 1st ed. 1957 |
| *Vasavadutta*, 1915 - 1916 | 1st ed. 1957 |
| *Urvasie*, 1893 - 1896 | 1st ed. 1896 |
| *Ahana and Other Poems*, 1895 - 1915 | 1st ed. 1915 |
| *Love and Death*, 1899 | 1st ed. 1921 |
| *The Viziers of Bassora*, 1893 - 1905 (Baroda) | 1st ed. 1959 |
| *Eric*, 1912 or 1913 | 1st ed. 1960 |
| *The Chariot of Jagannath*, 1918 (translated from the Bengali) | 1st ed. 1972 |

## 5. Political Period

| | |
|---|---|
| *The Ideal of the Karmayogin,* 'Karmayogin' 1909 - 1910 | 1st ed. 1918 |
| *The System of National Education,* 'Karmayogin' 1910 | 1st ed. 1921 |
| *The National Value of Art,* 'Karmayogin' 1909 | 1st ed. 1922 |
| *Speeches*, 1908 - 1909 | 1st ed. 1922 |
| *The Doctrine of Passive Resistance*, 1907 | 1st ed. 1948 |
| *Bankim - Tilak - Dayananda*, 1907 - 1916 - 1918 | 1st ed. 1940 |
| *The Brain of India,* 'Karmayogin' 1909 | 1st ed. 1921 |
| *Tales of Prison Life* (translated from the Bengali) | 1st ed. 1974 |

The complete works of Sri Aurobindo
30 volumes (1972)

# REFERENCES

*Most of the quotations from MOTHER'S AGENDA have been considerably abridged, at times composed of different extracts, otherwise several volumes would have been necessary. We apologize for this, but the readers of the AGENDA will have the joy of discovering for themselves the pure essence of the unabridged text. Most references to the works of Sri Aurobindo correspond to the original English of the Centenary Edition. Numbers in bold type indicate the volume number.*

**Chapter 1**
1. Savitri, 1.2.19

**Chapter 4**
1. Savitri, 1.4.55

**Chapter 6**
1. Thoughts and Aphorisms, **17**.125

**Chapter 7**
1. The Supramental Manifestation upon Earth, **16**,2,17

**Chapter 8**
1. Sri Aurobindo, *Unpublihed Letter*, Mother India (Sept. 1975)
2. Savitri, 2.12.278

**Chapter 10**
1. Savitri, 10.3.638

**Chapter 11**
1. Letters on Yoga, **22**.340

**Chapter 13**
1. White Roses (8/28/1966)
2. Thoughts and Aphorisms, **17**.145
3. Collected Poems, **5**.101

**Chapter 14**
1. Questions and Answers (5/20/1953)
2. Savitri, 3.3.317

**Chapter 15**
1. Savitri, 1.1.7

2. Questions and Answers (1930–31) Aphorismes et Paradoxes, p.46

**Chapter 17**
1. Questions and Answers (12/28/1950)

**Chapter 18**
1. White Roses (12/6/1965)
2. Savitri, 1.4.55

3. Mother India (April 1974)

**Chapter 20**
1. Savitri, 4.3.370
2. Savitri, 6.2.461
3. Ibid.
4. Savitri, 1.2.17
5. Savitri, 1.4.55
6. Savitri, 9.2.584
7. Savitri, 11.1.708